Electrical Interference in Electronic Systems:

Its Avoidance within High-Voltage Substations and Elsewhere

Electrical Interference in Electronic Systems:

Its Avoidance within High-Voltage Substations and Elsewhere

R.E. Martin, C.Eng., D.F.H.
F.I.E.E., F.I.E.R.E., Consultant

RESEARCH STUDIES PRESS

P.O. Box 92
Forest Grove, Oregon 97116

8 William Way
Letchworth, Herts
England

Electrical Interference in Electronic Systems

R.E. Martin, C.Eng., D.F.H.
F.I.E.E., F.I.E.R.E., Consultant

Printed in the United States of America

Foreword

This book is intended to be used in conjunction with other documents referenced to facilitate interference immunisation by isolation, screening, filtering, earthing and scheme design for use with electronic systems in electrically noisy environments.

The book particularly concerns the use of electronic equipment in high-voltage electricity substations and outlines the ways within an electricity station in which harmful interference may enter the electronic systems as well as indicating appropriate measures for the control of the effect of such interference. The book is not concerned with interference effects external to the station which are considered by C.C.I.T.T. and elsewhere. It also includes material especially relevant to the high-voltage station case which is intended as an addendum to the information given in the references. In addition, the book considers ways to obtain more data on the magnitude of the interference problem so that it would later become possible to quantify the extent of the attenuation required in practical cases. Means of setting the acceptance criteria are also considered in Chapter 8.

Conclusions, references and a bibliography are also included.

Although every effort has been made to ensure accuracy of the contents, neither the publishers, the authors nor the C.E.G.B. accept responsibility for errors or omissions, or for any consequences however caused which may arise from the use of material contained herein.

Contents

1 Introduction

Semiconductors used in modern electronic equipment and in computers are very sensitive to interference, and due to their low thermal inertia they are easily destroyed by overloads lasting even one millisecond. Many such devices are normally operating at microwatt or milliwatt power levels, so that even if destruction is avoided false operations can easily be caused by the ingress of undesired interference. A high-voltage substation is a prevalent source of intense interference, and the satisfactory use of semiconductors in such a station will necessitate the use of isolation, screening and filtering techniques.

Computers are one of the most difficult items to protect against interference because the spectrum of the signals they use is extremely wide. The pulses used for signalling within the computer may have durations of a few tens of nanoseconds and rise times of a few nanoseconds (1 nanosecond $= 10^{-9}$ second). They will thus contain frequencies extending to 100 MHz or more. Very short interference pulses can easily be interpreted by the computer as true signals, and this may corrupt a program in store or inhibit the running of a program. Furthermore the interference immunity levels of integrated circuits reaches a minimum valuc with pulses a few tens of nanoseconds long, emphasising the requirement for filtering of interference signals in the range extending to tens of MHz.

Because of the smaller physical sizes of the plant, the spectrum of interference due to switching and isolator operations tends to include higher frequencies when lower-voltage networks are considered. On the other hand, this is to some extent offset by the larger powers and increased stored energy available in the highest-voltage stations where the spectrum of interference tends to include lower frequencies because of the large physical size of the components forming the residual impedances and stray capacitances.

The extent of any measures will depend on the environment, coupling and susceptibility of the equipment being immunised. The measures cannot therefore be specified universally. Nevertheless, guidance is given in this book which, if used in conjunction with the documents listed in the References and Bibliography and when implemented correctly, can be expected to ensure high degrees of interference immunity. The precise statement of the ways in which a given type of interference can jeopardise the operation or even damage some device (such as a computer) is very difficult to determine because the magnitude of the parameters is often not known and the equivalent circuits are not only very complex but often cannot be defined. As it is usually not possible to quantify the precise conditions leading to an idealised solution, the only available course is generally to provide such measures as have been shown, by experience, to be effective in reducing the incidence of interference problems. Similarly, measurements of transients and electromagnetic fields made in substations often cannot be usefully applied in the design of preventive measures because the repeatability and mode of coupling are ill-defined. Nevertheless, measures are well known in sound engineering practice and have been evolved over the years as a result of experience by designers, manufacturers and system users. Some of the measures are less well known, and experience indicates that they also could often be employed with advantage, and at small cost, as a precaution against certain forms of interference which are prevalent in the hostile environment of a substation. The measures outlined in this book are designed to counter electromagnetic radio, audio and power interference produced by switching operations, faults and lightning. Damaging radiofrequency interference may extend to, say, 100 MHz, and its exclusion requires rigorous attention to elimination of earth loops, and provision of screening and filtering (Harrison, 1965, Cox, 1975). Such considerations apply not only to the substation equipment (for example, conventional cubicles do not provide sufficient screening) but also to the choice and design of electronic systems (return circuits within the electronics by separate unique connections for each section of the circuit to the chassis, use of high noise immunity logic, etc.).

Because of the wide frequency spectrum with which we are concerned, even short conductors must be considered as lines having distributed constants. An earth conductor 3.75 m long, for example, has a ¼ wave resonance at 20 MHz and acts as an excellent radiating antenna.

A single wire only 1 meter long has an inductive impedance of several ohms at MHz frequencies. One cannot therefore depend on the use of earth connections whatever their copper cross-section to derive an equi-potential. Cox (1975) has shown that interference currents flowing in cable screens can reach 100 mA at frequencies of up to 100 MHz (see also Section 10.11). A stray capacitance of only 15 pF has a reactance of 10 kohm at 1 MHz and corresponds to only some 15 cm of shielded cable or to a 1 meter length of parallel 0.015 in copper strip conductors printed 0.03 in apart.

Conventional practices in providing low-inductance earth connections are therefore unlikely alone to be a sufficient measure against interference; and, although essential at DC, 50 Hz and low radiofrequencies, must be supplemented by local filtering to remove high-frequency components and block all routes by which such interference can gain ingress to the vulnerable equipment. Nevertheless, a local reference earth point is clearly a requirement in respect of human safety where exposed metalwork is involved. However, it is often undesirable for this earth connection to form a part of the measurement or signal circuit.

The growth in use of mass-produced electronic equipment in substations to replace craftsmen-built electromechanical devices, coupled with the trend towards solid-state and integrated-circuit electronics, with their increased susceptibility to interference and damage, focuses attention on the need for

2 Scope of This Book

No book can define the degree of interference susceptibility reduction needed in each case, since this will depend on the noise environment which can change with time, the susceptibility of the equipment subjected to the interference and the significance of the correct operation of the equipment under test. Nevertheless, it is possible to indicate in general terms such measures, which if taken, whilst not guaranteeing immunity, can be expected to give acceptable operation in practical cases.

It should be noted that interference can often be troublesome by a variety of means of ingress rather than from a single cause. Therefore, omission of precautions may lead to difficulties being experienced.

To facilitate quick use of this book, Chapter 9 gives a brief check list of appropriate measures against interference. Material expanding on each aspect is included in separate chapters.

This book generally treats individually each one of several representative ways in which harmful interference can enter a system by first outlining the mechanisms and then indicating what measures should be taken to avoid the particular problem. Simplified schematics in a "before treatment" and "after treatment" sense are included, since diagrammatic representation is often much clearer and briefer than a descriptive text.

The book also considers interference immunity of computers and solid-state systems and makes specific recommendations in those cases where the designer has freedom of choice.

Where such freedom does not exist, as for example in the use of proprietary computer systems, recommendations on appropriate measures are given

The References and Bibliography in this book will be found helpful in the design of systems. Much information on interference measurement, pick-up and remedial measures is given in the literature. There are several classical treatments (Harrison, 1965) (Rhodes, 1966). The latter considered the computer-instrumentation interface.

Because so much has been written about interference, an attempt has been made in this book to minimise repetition by quoting appropriate references. For this reason, the serious reader is urged to make use both of the references and also of the codes of practice and other documents listed in Chapters 14 and 15. In particular the following are recommended:

a. Bull, J.H., "Code of practice for the avoidance of electrical interference in electronic instrumentation and systems," 1975,
b. U.S.A.F., "Instrumentation grounding and noise minimisation handbook," (U.S.A.F., 1965),
c. Morrison, R., "Grounding and Shielding techniques in instrumentation," Wiley, 1976,
d. Ficchi, "Electrical Interference," Iliffe, London, 1964
e. Ministry of Defence Ship Dept. (MOD), 1972, "Electromagnetic Compatibility Guide," TP1006,
f. Anders, R. and Campling, A.C., "Investigations into interference in substation and power station auxiliary cabling," (Anders and Campling, 1976),
g. Interference with Electronic Instrumentation and Systems," (Electrical Research Association, 1968).

References are also given in Chapter 15 to extensive published bibliographies, one of the most useful, as it contains abstracts, being that published by the British Steel Corporation (Healey, Mrs. A.M.). Reference is made by Schlicke (Schlicke, 1972) to a "Noise Guide" produced by U.S.A. industry and the I.E.E.E. There is also a "Radio Frequency Interference Handbook" of NASA (Taylor, 1971).

The MOD Guide (e. above) gives examples of interference calculations (Section 4.1.1.8).

In particular, the CIGRE paper by Anders and Campling (Anders and Campling, 1976) provides an excellent review of the mechanism of interference production and coupling from high-voltage plant.

A detailed study of the problem is also contained in the CIGRE Report 36.05, 1970 (Borgvall, Holmgren, *et al.*) and the discussion thereof. This reference contains results of measurements made in substations.

3 How to Use This Book

This book is intended to be used in conjunction with the papers given as references.

The contents list indicates the field covered.

Chapter 4 summarises each of the basic discrete means by which interference may gain ingress to a system having particular reference to the substation case. In practice, variations on, and combinations of, the basic means will occur.

Critical examination should be made of any proposed system *to check that all of the means described have been protected against.* This protection will comprise such shielding, filtering, signalling, earthing and supply arrangements as are necessary to achieve the desired interference immunity so as to permit continued normal operation of the equipment and to prevent damage and harmful effects.

Chapter 4 includes simplified typical schematics illustrating each type of interference ingress. Where appropriate, the diagrams are in two parts showing the circuit before and after the treatment necessary to control the particular variety of interference ingress.

Additional information and references to interference measurements in substations, etc. are given in Chapter 5.

Chapters 6 and 7 discuss the susceptibility of computers and of integrated circuits respectively to interference.

Chapters 8 and 9 consider the means for setting realistic attenuation levels to be specified in future and where possible give typical attenuations needed with a check list of desirable precautions.

Chapter 10 considers test specifications.

The conclusions and recommendations are included in Chapters 11 and 12.

Where specific types of hardware are mentioned, many examples of typical manufacturers are referenced to an address list given in Appendix A. It is emphasised that inclusion of manufacturers' or suppliers' names in Appendix A does not imply that the products concerned are necessarily approved or available, nor that other manufacturers may not offer comparable or superior alternatives.

3.1 Glossary of Terms

A comprehensive glossary of terms is given in an E.R.A. report (Bull, 1975). Anders and Campling (Anders and Campling, 1976) have defined very clearly the coupling modes for substation interference. A further addition to the glossary of terms is included below:

Vulnerable equipment

The electronic equipment to which protective measures are being applied so as to prevent malfunction and/or damage. Such equipment may be sensitive to *potential differences,* themselves often resulting from the flow of current through an impedance.

True earth

An equipotential earth whose potential is unaffected by the passage of transient or fault current and which therefore does not have an impedance in common to the passage of such currents. The potential of true earth can be seen as that of a potential measurement electrode at a remote point beyond the potential gradients surrounding any current carrying earth electrodes.

Peripheral system

All items of equipment electrically externally connected to the vulnerable equipment (q.v.) other than the power source. Peripheral systems include teletypewriters, V.D.U.s, remote A.D.C.s, modems, communication links, current and voltage transformers, auxiliary switches together with their wiring and cabling. From the interference pick-up viewpoint, they thus often constitute the most vulnerable part of the whole system.

Screening

Screening in the form of mesh or sheet metal enclosures (U.S.A. name "shielding"). Depending on the requirements and the screening design, the effect may be predominantly electric or magnetic. It becomes electromagnetic where one deals with electromagnetic radiation.

Station earth mesh

The substation earth bar. Owing to its extensive size, residual internal impedances are often significant at high frequencies and this can result in local potential rises because of the large transient and fault currents that can flow. The station earth mesh is connected to True Earth (q.v.) through an impedance which may be lower than the internal impedances. This implies that a mesh having small enclosed areas and multiple interconnections is advantageous.

Clean instrumentation earth

An earth system of minimum impedance, joined to the Station Earth at only one point and having minimum capacitance and mutual coupling to the local station earth.

Site of malfunction or damage

The site within the Vulnerable Equipment (q.v.) where a

malfunction or damage could occur in the absence of proper precautions.

Cell-type screening enclosure

Prefabricated screened enclosure of double-walled metal sheet or mesh construction.

Conducted interference

Interfering signals carried by a conductor. Conductive coupling exists when two circuits, which may have independent functions, are coupled by an impedance common to both circuits. The common impedance is often a common earth connection or a ground return (Anders and Campling, 1976).

Conductive gasket

Resilient metallic gasket used to reduce electromagnetic energy leakage at access openings in screening enclosures.

Integrated-circuit abbreviations (see Chapter 7)

DTL	Diode-transistor logic
TTL	Transistor-transistor logic
MECL	Metal oxide emitter coupled logic
HTL	High threshold logic
ECL	Emitter coupled logic
CMOS	Complementary metal oxide silicon
LPSTTL	Low-power Schottky transistor-transistor logic
EMC	Electromagnetic compatibility. The whole subject of interference immunisation.
RFI	Radiofrequency interference.

Contact finger

A conducting metal strip with fingers used in compression wiping seals around doors and access ports in screening enclosures.

Double-screened enclosure

A type of double-screened enclosure in which the inner screen is electrically connected to the outer screen at one point only.

Screened enclosure

An enclosure capable of reducing ingress (or egress) of interference.

Filter

Capacitor and/or inductor network used to reduce "conducted interference" (q.v.). Filters operate by presenting a high impedance to conducted interference and include a shunt or "drainage" path to present a low impedance whereby interference currents can be diverted harmlessly, often by presenting a path across the input or from each side of the input to earth.

Galvanic isolator

See barrier device (conductive isolator, ohmic isolator).

Barrier device

A device inserted in a signal or power path to prevent the flow of interference currents. Barriers may be formed by the use of relays, double wound transformers, the use of capacitively or inductively coupled carrier signals, by self-contained optical light emitting diode (LED) isolators or by optical links over a distance. Faraday screens to guard and intecept undesired electrical couplings across the barrier are often desirable. Barriers may be effective against series mode or common mode interference or some combination of the two, depending on their type.

4 Interference Ingress

The electronic system to be protected is considered here as a single vulnerable device. Interference can enter this device directly or by way of any electrical conductors linking the device to the "outside world." Such conductors are a prevalent source of difficulty in that they can pick up unwanted interference by both electric and magnetic means, conveying such interference directly into the electronic system. Coupling can arise from conductive (potential difference), inductive and capacitive effects, as has been well illustrated in the literature (Anders and Campling, 1976). Because physically separated points within the substation may be at widely differing potential, currents flow in the earth system which, together with those in the busbars, etc., give rise to induction. The local rise of potential in the earth system results in conductive coupling where common impedances occur. The importance has also been shown of achieving a close balance at high frequencies in long signal cable circuits (see Section 4.5.1), so as to counter the effects of capacitive coupling with the signal conductors. The conductors may comprise power-supply leads and signal circuits to the peripheral systems. By signal circuits is meant both input and output of the electronic system, whether analog or digital. In a computing system, for example, the ingress of interference voltages through common mode rise of earth potential (Harrison, 1965) of a (remote) mains neutral earth connection can give rise to difficulties. Similarly, connection of peripheral

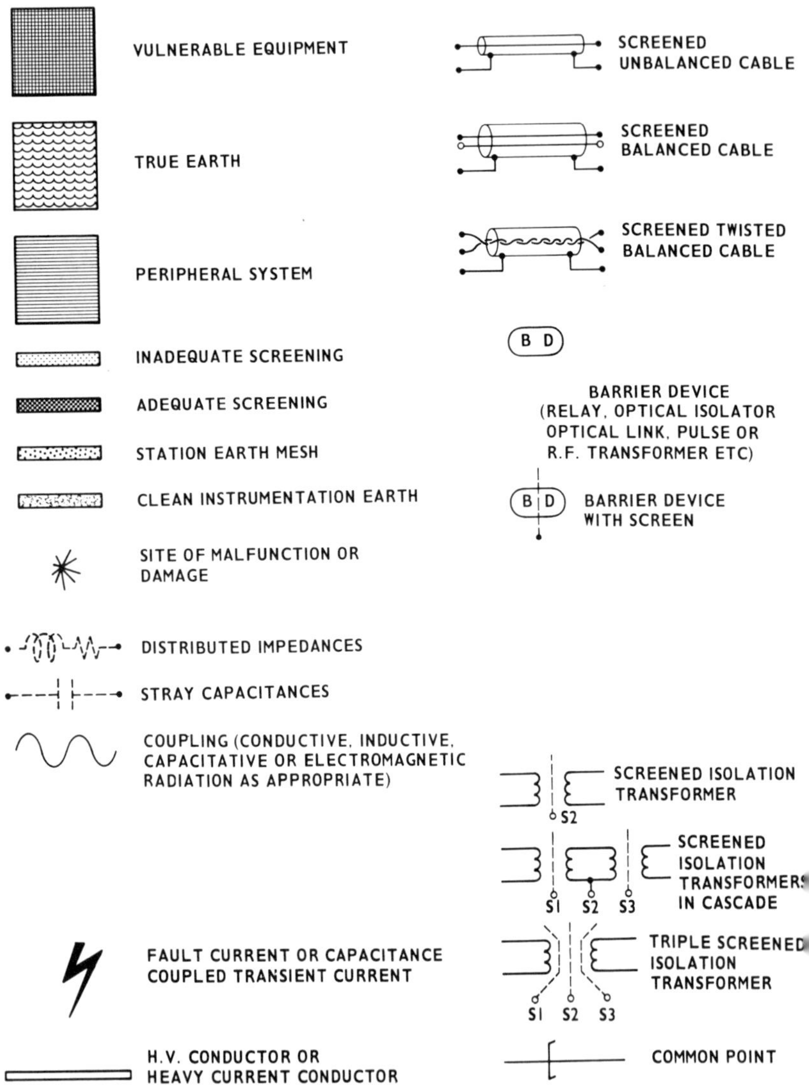

FIG. 1 KEY TO SYMBOLS USED

input-output devices such as substation voltage transformers, teletypewriters, visual display units and communication lines can also cause serious problems unless appropriate precautions are taken.

In many installations some of the peripheral devices will also themselves contain vulnerable electronic systems. They will also necessarily have to be treated. However, these do not differ from the basic case and to simplify this guide, only a single vulnerable device with a single peripheral device is considered.

Fig. 1 illustrates the symbols used throughout this book, and Fig. 2 illustrates the above simplified case.

Referring to Fig. 2, the possible means of ingress of interference are either:

(a) directly to the vulnerable system,
(b) via the power supply, or
(c) via the signal connections.

All of these must therefore be treated individually, and each can suffer from the causes enumerated in the following chapters.

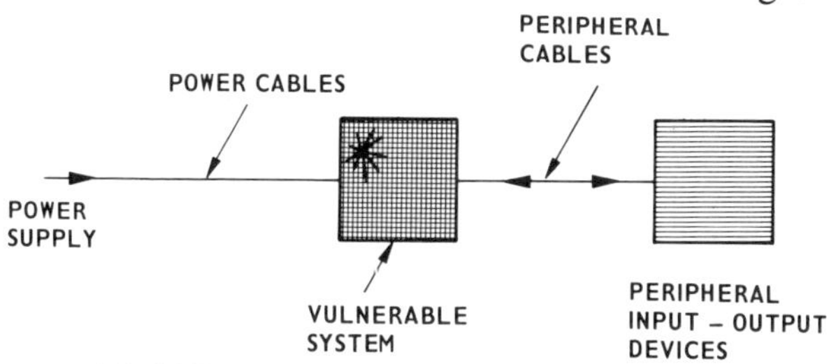

FIG. 2 BLOCK DIAGRAM ILLUSTRATING VULNERABLE SYSTEM

Fig. 3 illustrates and lists appropriate measures and explains diagrammatically the routes by which interference can be coupled. A detailed discussion of the mechanisms has been given elsewhere (Anders and Campling, 1976; Mainstrom, 1977).

Thorough screening and filtering of leads entering a computer employing pulses of a few tens of nanoseconds are essential. Fortunately, the response times required in respect of the substation plant may often be relatively slow (longer than 1 millisecond), so that it is inherently often unnecessary to use wideband input circuits. When the response time is chosen, the sensitivity to interference may often be very greatly reduced by using simple resistance-capacitance low-pass filters with a time constant of one or more milliseconds; in addition, protectors will also be used to avoid damage or maloperation resulting from the application of overvoltages.

Several of the line attenuation devices commonly used are more properly protection devices because in operating to prevent damage they may prohibit normal working of the system for a period. Spark gaps, arresters and gas-discharge devices fall into this category. These are nevertheless very valuable in limiting voltage or current under transient conditions (see Entries 13 and 43 in Appendix A). At lower power levels, Zener diodes are widely employed either alone or in conjunction with resistors to protect circuits from transients of 100 W to 30 kW for 100 microseconds. A Mullard application note (Mullard, 1971a, Mullard, 1971b) gives recommended circuits and means of choice of the appropriate type (see Entry 46 in Appendix A). Standoff voltage ratings of 3.9 to 62 volts are commonly available in the silicon surge suppressor diode range. The "transorb" devices are an alternative (see Entry 27 in Appendix A).

A range of modular terminal block elements containing level clamping Zener diodes are offered (see Entries 35 and 59, Appendix A).

Overvoltage protection modules for the protection of electronic loads operating off supplies in the range 5-28 volts and at currents up to 20 A are also obtainable (Entry 35, Appendix A). The clamping levels will generally be chosen to be well below the level at which false operation or damage can be caused to the vulnerable equipment (see Chapter 7).

Where D.C. digital signal circuits are involved, "D.C. crowbars" are used (Entry 53, Appendix A). Typical crowbars can switch in a few microseconds from an open circuit to a virtual short circuit. Recovery is achieved when the power is removed. Trip levels of 4.7 to 600 volts at current levels from 3 A to 235 A continuous are obtainable (Entry 41, Appendix A).

Attention is drawn to the desirability of reducing interference at source wherever possible. For example, power-switching contactors are readily fitted with appropriate suppressors at their control and output contacts (see Section 4.5.5). R.C. elements are offered for this application (Entry 38, Appendix A).

Where such nonlinear elements are connected from each side of a circuit to earth, it should be noted that conduction of only one element may convert a common mode interfering signal into a transverse (or series) mode. Unless precautions are taken to ensure simultaneity of conduction, damage may be

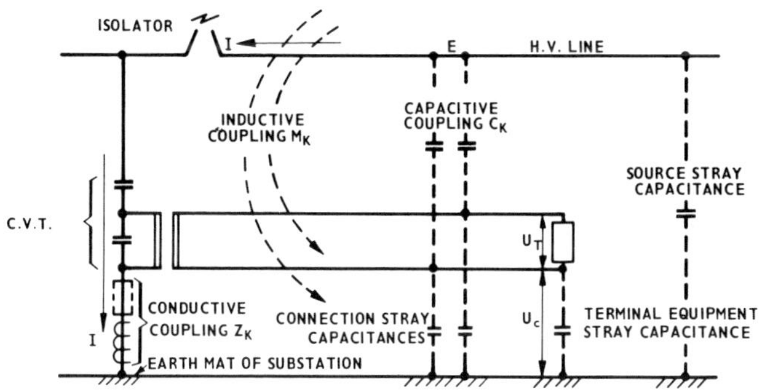

U_T – TRANSVERSE MODE VOLTAGE

U_c – COMMON MODE VOLTAGE

(AFTER ANDERS & CAMPLING, CIGRE PAPER 36 - 09, 1976 - REPRODUCED BY PERMISSION)

CONDUCTIVE	INDUCTIVE	CAPACITIVE	E.M. RADIATION
SYSTEM LAYOUT	SYSTEM LAYOUT	SYSTEM LAYOUT	SYSTEM LAYOUT
SEPARATION	SEPARATION	SEPARATION	SEPARATION
SINGLE POINT EARTHING	BALANCED SIGNALS	INTERCEPT BY SCREEN	REDUCTION AT SOURCE
VOLTAGE WITHSTAND	TWISTED PAIRS		SCREENING SOURCE
REDUCTION WIRES	SCREENING (MAGNETIC)		SCREENING VULNERABLE EQUIPMENT.
MINIMISE U_c	REDUCTION WIRES		FILTERING
GUARDING U_c			
SPARK GAPS	SPARK GAPS	EARTH PLANES	ELECTROMAGNETIC SCREENING
DISCHARGE TUBES	DISCHARGE TUBES	COUNTERPOISE	GASKETING OF DOORS
DIODES	DIODES	CIRCUIT EARTHING	INTEGRITY OF SCREEN
REDUCTION WIRES	MAGNETIC CABLE SHEATHS.	SCREENING (ELECTRIC)	FILTERING CONDUCTORS ENTERING
NEUTRALISING TRANSFORMERS.	BALANCE OF SOURCE AND SINK.	CABLE SHEATHS	NO APERTURES
TRANSFORMERS	TWISTING OF CABLE CORES.	REDUCTION OF CIRCUIT IMPEDANCES.	
OPTO-COUPLERS			
RELAYS			
OPTICAL FIBRES			
CHOKES			
FERRITE RINGS			

FIG. 3

EXAMPLE OF CONDUCTIVE, INDUCTIVE AND CAPACITIVE COUPLINGS AND SOME MEASURES AND DEVICES TO COMBAT THEM

caused by the resultant series mode interference being directly applied to the measurement equipment.

Normally the protection elements will be followed by a barrier device and/or radio-interference suppressor after which the cables will be within the clean area, i.e., within the screened enclosure.

Knox (1973) has given information on protector characteristics, and his paper is helpful in design of any particular system. The literature refers to transverse transient overvoltages in protecting secondary wiring (Wiszniewski and Marczonek, 1975).

Generally speaking, the interference ingress situation in substations differs somewhat from that in power stations. In a substation the reduction of common mode interference and of electromagnetic interference presents the major problems. In a power station, however, the more widespread use of low-level signals gives rise to series mode interference problems.

4.1 Interference

Operation of high-voltage isolators and switchgear, the switching of inductive loads, fault currents, etc., gives rise to intense broadband electric and magnetic fields, as well as potential differences. (Teleprotection, CIGRE). Interference can be coupled to all parts of the system illustrated in Fig. 2; and depending on their susceptibilities both the vulnerable system, the cabling and the peripheral devices may all need to be treated.

Genuine electromagnetic radiation also arises at the higher frequences, and this is dealt with by the normal radiofrequency screening methods (Sections 4.1.1.1, etc.) The remedial measures required differ depending on the form of interference. For example, radiofrequency shielding by a thin copper box may well be useless where low-frequency magnetic fields are the cause of malfunction. Similarly, screening of vulnerable equipment may be unnecessary if the interference gains ingress due to lack of common mode barriers against rise of local earth potential on an external cable system.

As an indication of the order of the magnetic 50 Hz field strengths likely to be encountered, a peak of 25.5 A/m was measured during a system fault in the 275 kV substation at Harker at a distance of 7.5 m from a conductor carrying 12 kA peak. Barratt (1965) reported peak fields of 37.5 A/m at the transformer resonant frequency 5 m distant from busbars connected to a 275 kV transformer. The field strength is proportional to current and decreases almost in inverse proportion to distance. Insofar as the vulnerable equipment itself is concerned, there is thus every advantage in locating it well away from heavy-current conductors, including those parts of the station earth mesh and cable sheaths likely to carry such currents. However, cables connecting to transducers or other peripheral equipment may have to be run through regions where intense fields can occur, and appropriate measures to minimise induction and prevent the flow of induced interference currents are necessary (see Sections 4.1.2 and 4.5). The method of measurement of electromagnetic fields is considered in Chapter 5. The susceptibility of computers to electromagnetic fields is examined in Chapter 6, but it is sufficient at this point to note that high-frequency fields of the order of 6 A/m peak can cause malfunction of a typical computer. Thus, if such a computer were installed in an environment where H.F. fields of 25 A/m were expected, screening arrangements having an attentuation capable of reducing the field within the computer to below 6 A/m would be essential.

Laboratoire Belge de l'Industrie Electrique, Rhode St. Genese 1640 Belgium (Laborelec) have considered a worst-case 50 Hz magnetic field to be that likely to occur under conditions of a single-phase earth fault on the first transmission line tower, the line passing right over the building containing the computer. In their system this corresponds to a 50 kA rms current, and this is taken to produce a magnetic field of 500 A rms/m at 50 Hz at the computer.

In many devices, such as visual display units (VDU), contact-making keys at the operator interface are sometimes being replaced with Hall-effect devices where depression of a key moves a magnet to electronically switch the Hall-element. Caution will need to be taken in the use of such devices in the substation environment as excessive environmental magnetic fields could cause false keyboard entries to occur.

4.1.1 Screened rooms or enclosures

4.1.1.1 Screening, general

Screening of the electric field is achieved by use of a conducting screen. Screening of the magnetic component of the electromagnetic field results from the use of a magnetic (steel, iron, mumetal) screen or a conducting (copper or aluminum) screen. Screening effectiveness is limited by discontinuities and apertures, and their effect becomes more significant at the higher frequencies. The screening effectiveness is a function of the permeability and conductivity of the screen material, and its thickness as well as the shape and size of the screen. The screen operates by the two effects of reflection from the metal and by the attenuation of a field passing through the metal (Fig. 4). A U.K. Ministry of Defence publication (MOD, 1968) explains this effect. Usually the electric screening effectiveness is greater than that of the magnetic screening, particularly for non-magnetic materials.

4.1.1.2 Size of screened enclosure, single or double screening

Clearly any screened enclosure in attenuating radiated electromagnetic fields can support natural internal resonances the frequency of which will depend, for example, on the

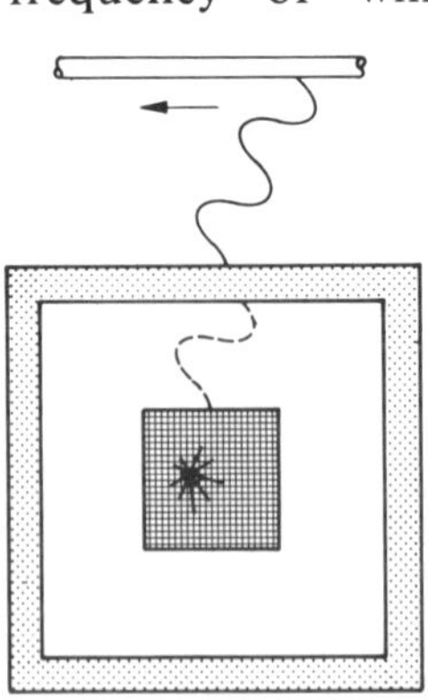

SINGLE INADEQUATE ENCLOSURE

FIG. 4

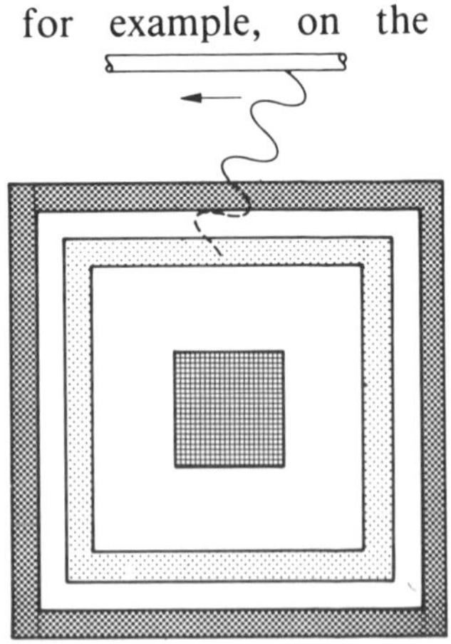

DOUBLE ENCLOSURE

FIG. 5

ELECTROMAGNETIC SCREENING

capacitances between opposite walls and the inductances presented by the other two walls. At the natural resonant frequency, currents flowing in parts of the walls cause reradiation and the enclosure thus becomes transparent at the particular frequency or frequencies. Consequently, where attenuation extending to high radio frequencies is a requirement, reduction of the size (and volume) of a screening enclosure (so as to place these resonance frequencies beyond the required range) is useful. This may often be achieved by subdividing a screened enclosure into several smaller enclosures. However, leads passing through the partitions may then need to be filtered to prevent the subdividing wall from being bypassed. Another means of improving screening effectiveness against radiated electromagnetic fields is to employ two or more enclosures one inside the other so that reradiation resulting from the flow of induced currents in the wall of one enclosure is further attenuated by another enclosure (Fig. 5). Such enclosures *must be electrically insulated one from another and bonded at only one point.*

As schemes in substations invariably utilise external connections, Figs. 4 and 5 are somewhat academic but are included to illustrate the point that even if the external connections could be discounted (e.g., by perfect filtering) there may still be a requirement to screen the vulnerable apparatus itself against electric fields, against magnetic fields and perhaps also against true radiated electromagnetic energy. Fig. 6 applies to screening provided to combat radiated electromagnetic fields.

It will be realised that it is important to establish in any given case the main function of the screening, namely whether the requirement is to combat electric fields, magnetic fields or true electromagnetic radiation, and to distinguish clearly between the different cases.

4.1.1.3 Earthing

Earthing of a screened room has no effect inasmuch as the electromagnetic interference case under consideration is concerned, but it is of course necessary in respect of safety and also, very importantly in a substation, in respect of electrostatic screening (Section 5.3).

Only one connection should be made from the outer screen to earth. If more than one connection is made, galvanic circuits to the station earth mesh may exist whereby some part of a fault or transient current may flow to earth via the screening. Such currents flowing within the screening material can of course give rise to radiation within the enclosure and destroy the benefit of the enclosure (Fig. 7).

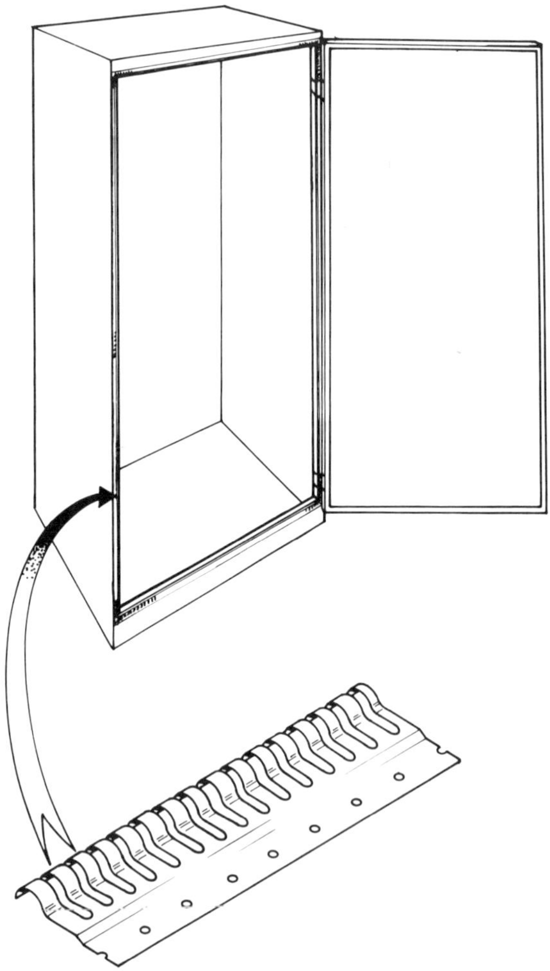

FIG. 6 CONTACT FINGER STRIP FOR R.I.F. SEALING OF DOORS

To avoid this problem in respect of cable sheath currents, it is usually necessary to arrange for *all* cables entering the enclosure to do so through a *single* gland plate. This may also incorporate barrier and filter devices contiguous with one of the enclosure walls, this plate also having a single low-inductance connection to the station earth (Figs. 8 and 9). A path still re-

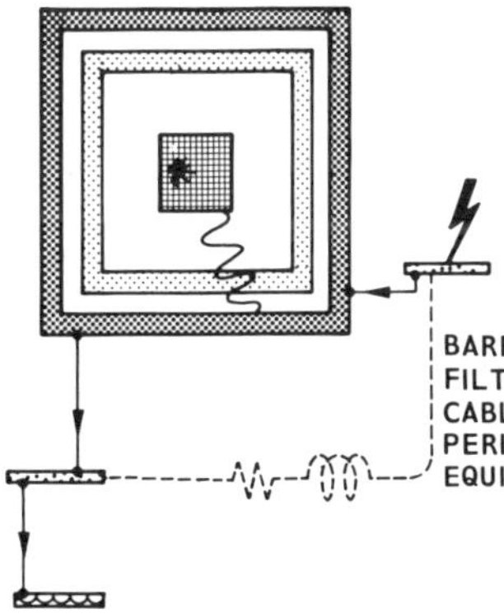

FIG. 7 MULTIPLE EARTHS TO SCREENED ENCLOSURE RESULT IN INTERNAL RADIATION

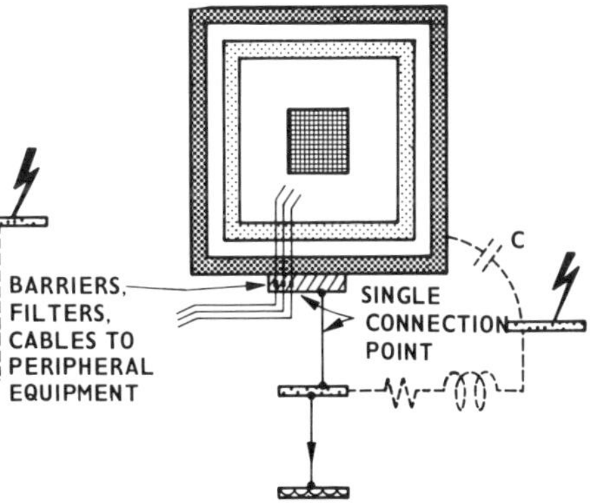

FIG. 8 SINGLE EARTH POINT ON SCREENED ENCLOSURE WALL MINIMISES INTERNAL RADIATION

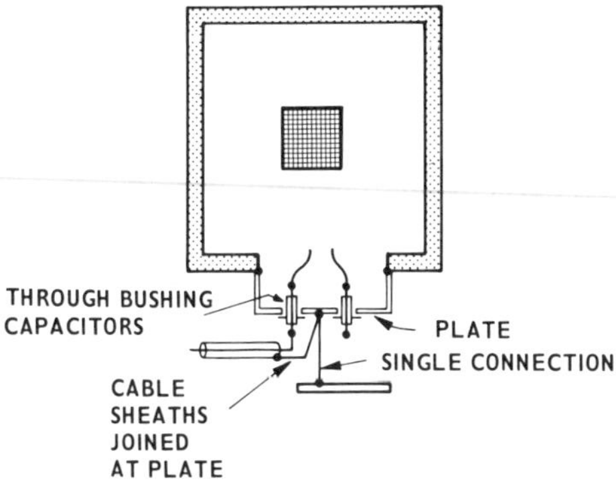

FIG. 9 MOUNTING FILTER CAPACITORS

mains via the stray capacitance C, and this should be minimised where construction considerations permit.

Fig. 10 shows diagrammatically the entry of conductors into an enclosure, the plate forming a part of the enclosure. Where it is necessary to return cable sheath currents to a separate point on the local earth mesh (Section 4.6.5), the arrangement of Fig. 11 is useful whereby the plate is electrically insulated from the cubicle. However, as interference currents will flow in the plate radiation will occur above it. A second filter support plate bonded to the cubicle as in Fig. 10 (not illustrated in Fig. 11) is therefore a requirement in such cases.

4.1.1.4 Bonding, sealing doors against radiation

The performance of screened enclosures depends not only on the materials used but also on the means of bonding the various sections. When screening against radiated electromagnetic energy, it is important that joints between panels are made (and kept) continuous along the whole length or failing this by discrete bonds placed at frequent intervals along the length. Joints in mesh are more easily made than in perforated sheet material. A higher attenuation often results from using a double screen of inferior material than a single screen of superior material.

Doors are a prevalent source of leakage. Unless they are arranged to close onto a metallic spring finger contact all round the edges or to seat on a metallic braid gasket all round, serious leakage can occur (Fig. 6). There are specialist manufacturers of such materials known as Altoflex and Xecon (Entries 2, 9, 14, 23, 36, 58, 64, 65, 71, Appendix A).

Wire-mesh-screened enclosures are characterised by a low attenuation in the lower-frequency range and a decline in attenuation occurs above about 20 MHz. Metal sheet or foil enclosures are more effective, and only a thin foil of 0.1 mm thickness is adequate where electromagnetic radiation is concerned. Typical enclosures provide attenuations of 60 dB at 100 kHz rising to 100 dB over the range 1 MHz to 100 MHz (Miedzinski and Pearce, 1949; Siemens, 1969/1970). However, thin foil is useless against 50 Hz magnetic fields, thus emphasising

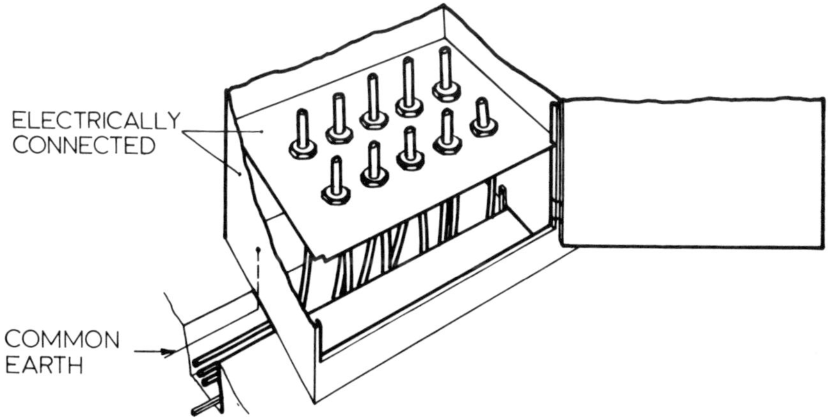

FIG.10 CONDUCTOR ENTRY THROUGH FILTERS IN COMMON PLATE BONDED TO CUBICLE.

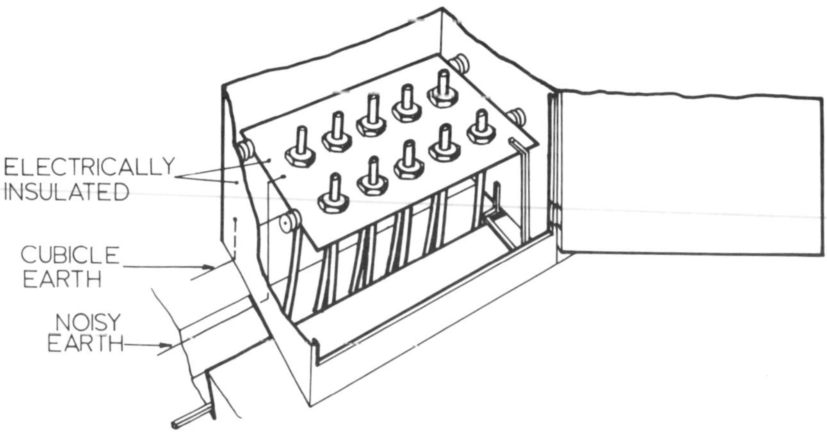

FIG.11 CONDUCTOR ENTRY THROUGH FILTERS IN COMMON PLATE INSULATED FROM CUBICLE AND RETURNED TO NOISY EARTH.

the need to establish the main requirement of any given screen.

The design of electromagnetic field screening enclosures requires special care in jointing the edges of sections and in sealing. Any untreated apertures, even if small, can drastically reduce the attenuation and can sometimes lead to a situation worse than that pertaining in the total absence of a screen. The provision of ventilation apertures necessitates the use of waveguides operating below their cutoff frequency (Siemens, 1969/1970). Manufacturers of electromagnetic field screening enclosures, such as Belling-Lee and Siemens, have developed modules from which screened rooms of known performance can be constructed (Entries 9 and 58, Appendix A).

4.1.1.5 Choice of material and construction

Workers at the Electrical Research Association have shown how to calculate the performance of a screening enclosure and have investigated the best materials to use (Miedzinski, 1959, and Miedzinski and Pearce, 1949). They have shown that high-conductivity, low-permeability materials are best for a screen in the form of a mesh. A closer mesh, even though this uses smaller-diameter wire, gives better results at high frequencies, but as stated already a metal sheet or foil is superior. Many of their findings are incorporated in the design of commercial screened rooms.

Any power supply or other cables (including telephone, lighting and heating cables) entering the enclosure must be filtered and/or screened to maintain this attenuation.

Where proprietary enclosures cannot be used, it is recommended that the papers by Sladek and the references cited therein should be studied (Sladek, 1966).

Sladek considers the basic principles and deals separately with the electric (or high-impedance) field and the magnetic (or low-impedance) field. Sladek gives the fundamental considerations for the propertics of screening (USA "shielding") materials and includes two tables. The first shows the penetration loss for a single solid metal screen of either copper or iron at various frequences from 60 Hz to 10 GHz. The second similarly shows the reflection loss for the electric field, the magnetic field and a plane wave.

Similarly the relevant literature (Shelkunoff, 1938; Shelkunoff, 1943; Miedzinski and Pearce, 1949; and Miedzinski, 1959) should be consulted. The latter papers give specific design information on screened enclosures. In most practical enclosures, the electric screening effectiveness is greater than the magnetic screening effectiveness (MOD, 1968, and App. 10, 5[a][c] thereof).

4.1.1.6 Inadequate interference precautions in commercial equipment

Fig. 4 represents diagrammatically the vulnerable equipment enclosed in an inadequate screen. Many well-known makes of small computers, for example, because they are designed for use in an environment other than high-voltage substations, have quite inadequate screening. At least one type has a plastic front panel to the enclosure so that the screening properties are drastically reduced. In addition, many such systems, even if they are well enough enclosed, have the potential benefit of the enclosure reduced by the passage of unfiltered connections from outside.

Reports of work indicate, for example, that the screening of a certain type of computer reduces externally applied electromagnetic fields to $\frac{1}{3}$ of their value; whereas in another type, because of improved screening, such fields are reduced to up to $\frac{1}{66}$ of their value, depending on the frequency.

Chapter 6 contains information on the susceptibility of computers to magnetic fields and illustrates the variation in screening efficiency as between different types of computer. Careful choice of the type of enclosure used for the vulnerable equipment, in accordance with the information in the guide, is a prerequisite. It will often be essential to enclose the vulnerable equipment (which may, as in Fig. 4, already be contained in an inadequate enclosure) in a second external and more adequate screening case (Fig. 5).

Experience has also shown that the use of cadmium-plated and passivated chassis, secured by screws without lock washers, can lead to imperfections in the screening against radiated interference, due to bad contacts. To secure radiofrequency screening, closely spaced screws with star-type lock washers are essential. Even small gaps or holes in the screen can result in an increase of interference susceptibility at certain frequencies; so great care is necessary that the surface is everywhere electrically continuous and that the screening effectiveness is not reduced by subsequent maintenance operations where failure to replace screws properly or improper coating of contact surfaces with paint can be detrimental.

In many commercial equipments, filtering of incoming connections is absent or is rendered ineffective by lack of screening or by looming filtered and unfiltered wires together. The availability of economical filters and filter-type connectors (see Section 4.5.5) should reduce this problem.

4.1.1.7 Practical screening in substations

Where electronic equipment is used in substations, it is therefore recommended that the appropriate precautions are taken in the design of the screen, and if this is difficult an external enclosure should be provided. Providing that the requirements are incorporated in the specification of the systems, the screening efficacy of the conventional enclosures of the equipment could often be improved. For example, electronic equipment is sometimes housed in metal cases with glass windows. A fully metallic housing is preferable; attention being drawn in cases where screening against radiated radiofrequency interference is a requirement to the need for a *continuous* bond all round any removable covers or doors. Where inspection windows are necessary, a metallic gauze behind the glass window enables continuity of the screening enclosure to be maintained.

Similarly, control cubicles already provide some measure of screening, but the use of either a metallic gasket (see Entries 2, 14, 71, Appendix A) or beryllium-copper or phosphor-bronze spring finger contact strip all round the doors could greatly enhance the attenuation (Fig. 5). *The bottom of cubicles should be screened and any earth strip run externally* (see Figs. 12, 13, 14, 15). (It is presupposed that cable cores entering a cubicle will be properly filtered—see Section 4.5.1.)

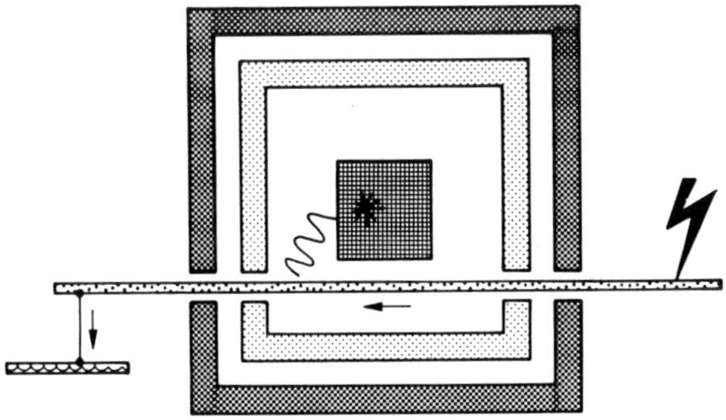

FIG. 12 LOCATION OF EARTH STRIP WITHIN ENCLOSURE CAUSES INTERNAL RADIATION

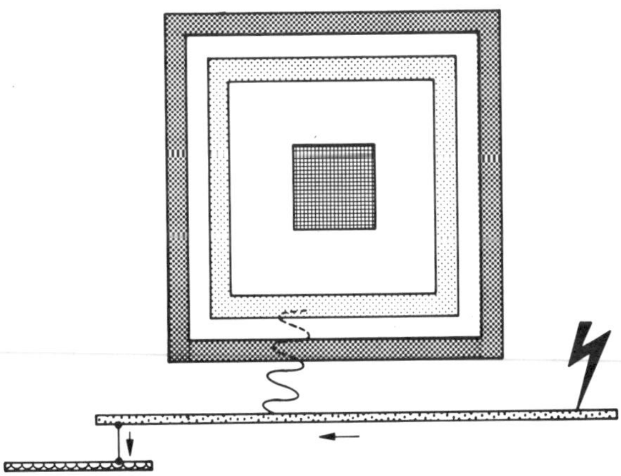

FIG. 13 LOCATION OF EARTH STRIP OUTSIDE ENCLOSURE AVOIDS INTERNAL RADIATION

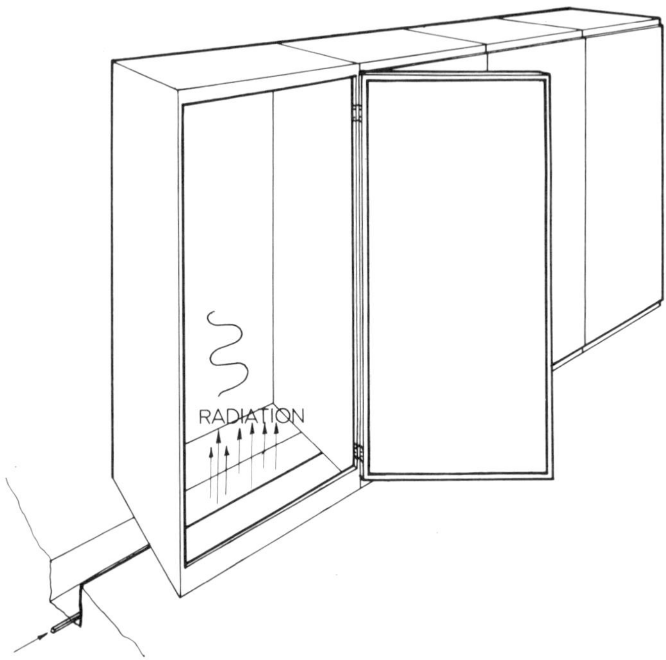

FIG.14 RADIATION FROM EARTH STRIP AND CABLE SHEATHS ENTERS ENCLOSUR

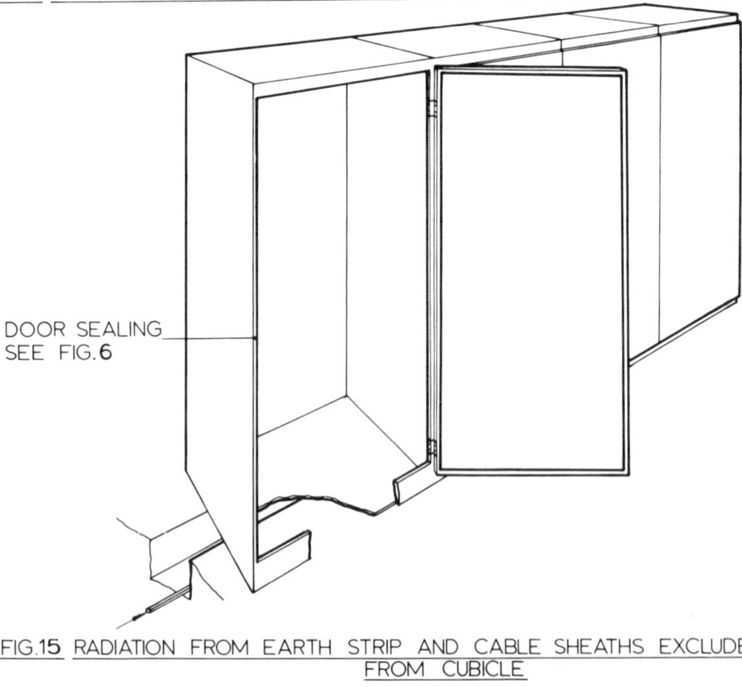

FIG.15 RADIATION FROM EARTH STRIP AND CABLE SHEATHS EXCLUDED FROM CUBICLE

Where sensitive electronic equipment is used, the supporting cubicle should preferably *fully* surround the device in its own containment, therefore providing the double screen of Fig. 5. No windows or apertures, unless closed with mesh connected at all edges, are permissible. Where this is not possible, an external single- or double-screened enclosure will often be essential.

As many electronic equipments in substations will be mounted in conventional cubicles, it may be possible to enhance the screening properties of the traditional cubicle by improvements to bonding and gasketing which could readily be incorporated.

Attention is drawn to the relatively high radiofrequency interference levels extending to many MHz produced by fluorescent lamps. In cases where highly sensitive equipment is mounted in the vicinity (antennas for radio receivers, for example), screening by enclosure of the lamps and provision of filtering at the lamp fittings may be needed. Where improvement of antenna siting is impossible, a transparent conducting film may be incorporated in the glass of the fitting to complete the screening enclosure, or an edge-bonded mesh may be built over the aperture.

4.1.1.8 Calculation of interference and screening

While some information exists on conducted interference (Turner, 1968), the information of its spectrum and that of electromagnetic fields due to switching operations, etc. is less adequate. Whilst theoretically it is possible to calculate the effects, any attempt to do so is likely to be unrewarding because the system parameters are usually not well enough defined to permit any confidence in the result.

One could consider worst-case conditions applicable to some given substation configuration by making large simplifications of, say, a busbar of stated capacitance charged to a maximum d.c. potential and being suddenly discharged into a Z_0 of the requisite value to derive the peak current and hence the electromagnetic field strength. Because one side of an isolator may be fed from the network and the other side may be intermittently coupled through a restriking discharge, the capacitance of the busbar can be charged to a d.c. potential of opposite polarity to the network potential so that when the discharge is re-established double the normal voltage can drive current into the circuit. In practice the often unknown stray capacitances and inductances with the effects of mismatch and travelling waves make the detailed modelling of a practical case more difficult because the circuit parameters are usually not known. For these reasons, a simplified worst case is considered to be adequate, until spectrum and level measurements are available (Chapter 5).

The reader concerned with the basic definitions and formulae is recommended to study the MOD Electromagnetic Compatibility Guide (MOD, 1968). This Reference contains Appendices which consider:

- Electric fields
- Magnetic fields
- Maxwell's equations
- Fields surrounding single isolated conductors, balanced twin conductors, coaxial cables, twin cables, balanced 3-wire systems and transmission lines
- Examples of magnetic interference calculations in twin cables, and multicore cables
- The wave theory of screening (transmission and reflection loss)
- Skin effect, including actual values
- Magnetic screening (nonmagnetic and steel enclosures)
- Field leakage and crosstalk in coaxial cables

Examples are given of magnetic interference calculations for two twin silicone-insulated cables laid side by side and for a 30-core silicone insulated layered multicore cable.

To avoid repetition, this fundamental material is not restated here, but the Reference is strongly commended, not only for the subjects listed above but also in relation to the whole topic of electromagnetic-compatibility measures (EMC).

In the U.S.A., the topics of Electromagnetic Interference (EMI) and Electromagnetic Compatibility have been treated together so that the grounding and shielding problems are often related to compatibility in a spectrum-utilisation sense which embraces radiofrequency assignment and use. A well known series of comprehensive seminars on "EMC—Design and Measurement for Control of EMI" has been established by D. White Consultants, Inc.

D. White Consultants have a range of publications on EMI noise, specifications, test methods, control methods, instruments and prediction techniques. They have devised a large number of computer programs which can be supplied as cards for use in programmable pocket calculators such as HP-65 or as decks of Fortran cards for use in IBM systems. Of possible interest to the power engineer, in relation to this book, are some programs of which a few are listed as follows:

1501 (605)	Building attenuation
1510	Shielding effectiveness, reflection loss
1511	Shielding effectiveness, reflection and absorption
1513	Total shield effectiveness
1520 (606)	Shielding effectiveness, cabinet or rack frame
1530 (608)	Shielding effectiveness, screen wire and holes
1601	Common impedance coupling
1401 (302)	Cable-to-cable coupling
1401 (304)	Cable-to-cable decoupling twists
1411 (305)	Cable-to-cable decoupling shield, loss tubing
1420 (604)	Common-mode ambient to-cable coupling, C-M rejection
1630	Ground loop coupling and susceptibility 1/N ratio

It is clear from examining the full list of programs from D. White Consultants, Inc. that the quantification of interference effects and the more scientific design of countermeasures have received increasing attention in the U.S.A., mainly as a result of military requirements. It is likely that much of this work is relevant to the subject of this book. D. White Consultants, Inc. have a 4,000-page, 6-volume series on EMI/EMC which also appears highly relevant. Their address is 14800 Springfield Road, Germantown, Maryland 20767, USA (Telephone:

301-948-0028, Telex 89-2766 DWCI GTWN).

Roessler, Amos, *et al.* (see Chapter 15) have given a formula for calculating the screening effect of copper (neglecting reflection loss):

$$dB = 3.34t\sqrt{f}$$

where t is the thickness of copper in thousandths of an inch
f is in MHz

For other metals,

$$dB = 3.34t\sqrt{f\sigma_r\mu_r}$$

where σ_r is conductivity of other metal relative to copper
μ_r is permeability relative to copper.

4.1.2 Cables

4.1.2.1 General

Cores in cables linking the separated items of control or instrumentation circuits are particularly susceptible to interference because long lengths are generally involved and these may traverse electromagnetically hostile areas. Screening is therefore required on instrumentation cables, and such a screen can be regarded as an extension of the screened enclosures at either end of the cable (Metcalfe, 1973, 1975).

Instrumentation cables should, wherever possible, be run at right angles to the main path of power and transient frequency currents whether in high-voltage busbars or cables. Except where the principal concern is to reduce the common mode voltages induced on the cores within instrumentation cables, these should not be laid parallel to earth continuity conductors. Cables should take the shortest route having regard to these considerations, so as to reach relatively clean areas where sensitive equipment is located.

Where power and signal cables have to follow the same route, e.g. parallel runs along the only available cable tunnel, the two types of cable must be kept separate by at least 0.3 meter and must not share the same support trays. Bull (1975) recommends a minimum spacing of 0.6 m with medium-voltage cables and 1.2 m with high-voltage cables. The level of interference induced in signal cables will not necessarily fall proportionally with increasing separation, as currents induced in cable support metalwork also contribute to the level of induction.

The literature refers to inadequacies of screening and bonding in power cables acting as a source of interference with instrumentation cables (Fowler and Gregory, 1970). This emphasises that the design of instrumentation systems depends to an extent on the interference sources. It may be difficult to reduce interference at source on, say, 400 kV busbars, but attention to layout, termination and the adequacy of screening of all power cables will be helpful in reducing interference.

4.1.2.2 Interference voltages induced in cables

Before interference in cables can be assessed, it is necessary to define the circuits and the connection of the cables as these determine the relevance of the various voltages which may be induced.

The preferred way of interconnection for most of the instrumentation and control circuits in substations is to use twisted pair cables so that balanced signal systems can be adopted. Then the two voltages of relevance are indicated in Figs. 16 and 17.

(a) Fig. 16 shows the voltage difference between cores. This appears as a voltage source in series with the circuit (transverse or series mode interference).
(b) Fig. 17 shows the voltage to earth appearing at one end of the cores. This will be approximately the same on each of the two cores forming the circuit. It is generally referred to as common mode interference. In this case it should be noted that the cable screen is now introduced into the circuit, giving rise to susceptibility to conductive interference (Fig. 3).

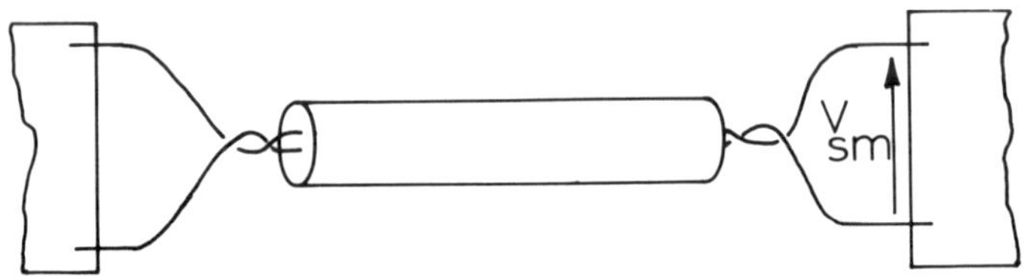

(i)

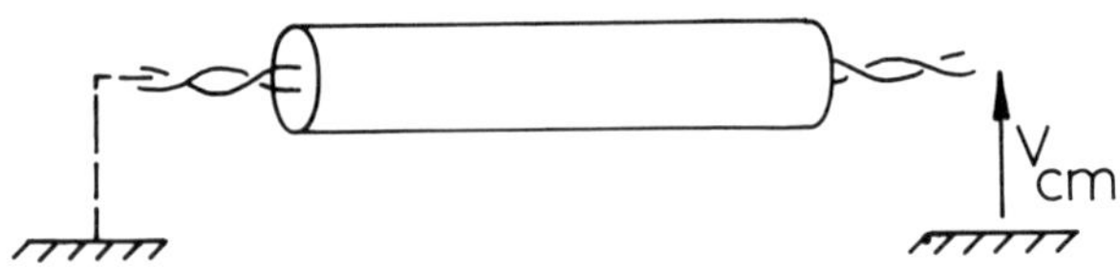

(ii)

FIG.16 CLASSIFICATION OF VOLTAGES INDUCED IN MULTICORE INSTRUMENTATION CABLES

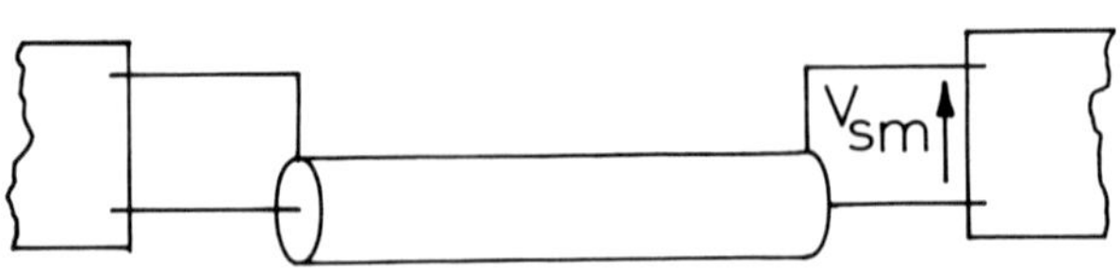

FIG.17 INTERFERENCE VOLTAGE IN CO-AXIAL CABLE SYSTEMS

For circuits which operate at high frequencies, it will be necessary to employ coaxial cables. Then the voltage between the core and the connection to the inner surface of the screen, Fig. 17, appears as series mode interference. The nearest equivalent to the common mode voltage referred to for twisted pair cables is the voltage between the outer surface of the screen and earth. The behaviour of coaxial and multicore cables should be considered quite separately as far as interference is concerned.

Interference voltages can be produced by coupling between the susceptible cores and conductors external to the cable, or with other cores in the same cable. It is external sources of interference which are mainly considered in this book, as it is in these circumstances that the highest power levels are found. A detailed consideration of the mechanism of conductive, inductive and capacitive couplings with particular reference to electricity substations is given in the literature (Anders and Campling, 1976).

Appendix B, comprising material kindly provided by E.P. Fowler of UKAEA, considers the mechanisms of interference coupling with cables and the interactions with electromagnetic radiation. Section B5 of Appendix B gives proposals for further work aimed at the provision of practical rules for use in given cases.

To avoid problems of interference due to internal coupling within the same cable, it is recommended that low and high level signals should not be included in the same cable (e.g., 100 millivolt transducer outputs are not compatible with 110 volt voltage transformer outputs). If the connection of two very different signal levels into the same cable is unavoidable, it is then preferable to use either multicore cables with unit construction and segregate the circuits with different levels into their own units, or else use cables with individually screened pairs (MOD, 1968).

J.R. Martin (C.E.G.B., T.D.C.D.) in a private communication dated 11 February 1976 describes experiments carried out at Norwich 400 kV substation to investigate the effects of cable screening during isolator operations. The dominant frequencies were found to be 400 kHz and 1.7 MHz, and an estimate of the energy being switched gave a result of 71.5 joules/phase. The report recommends the use of solid screened cables in addition to steel wire armouring.

Valuable guidance on the ways in which interference in instrumentation systems can be reduced is given in a paper by Wilson (1975). The paper refers to the importance of circuit layout in segregating screened and unscreened wiring and gives figures of typical values.

4.1.2.3 Cable screens and their earthing

As with screening enclosures, the effectiveness of a cable screen is a function of the permeability and conductivity of the materials used and its thickness, and the frequency of the interference. The need to have cables flexible for handling during their installation means that frequently the screens are constructed from woven braiding or interleaved tapes. A similar construction is adopted for the armouring, which is applied principally to provide mechanical protection, but which also acts partly as a screen. The discrete nature of this type of construction means that the screens are less effective than equivalent ones in the form of a homogeneous layer. Thus, in general, extruded screens or those formed by an overlapping longitudinally seamed tape are preferable to braided screens, and tape is preferable to wire for armouring. Wire armouring has no noticeable effect on screening efficiency (magnetic field attenuation). The efficiency of a cable screen can be increased by adding a second screen or mumetal tapes or an outer layer of armouring. The screening efficiency of multiscreened cables has been considered (Fowler, 1970).

The leakage and radiation of energy from cables have been applied usefully in communication systems on railways and in mine tunnels, and this has led to a considerable study of the means of predicting and controlling such leakage (Martin, 1975; Cree and Giles, 1975). The fundamentals of the leakage mechanism of leaky "feeder" cables has been examined (Martin, 1975), and this is helpful in enabling screening efficiency of different cables to be compared.

With cables which contain individually screened cores,

these screens should preferably be connected to the terminal equipment enclosure by means of a connector which maintains the symmetry of the screen rather than using a separate individual connection (Fowler, 1974).

The level of series mode voltage induced in a multicore cable by a field produced externally to the cable is reduced by a cable screen. At low frequencies (50 Hz), little reduction can be achieved without resorting to thick extruded screens or multilayer tapes of high-permeability material. Therefore the most effective technique for minimising low-frequency series mode interference is to use twisted pair cables and to maintain a circuit balance with respect to earth.

A cable screen can also reduce the level of common mode interference. This reduction is a result of the coupling between the cores and currents induced in the screen. A voltage is thereby induced in the circuit opposing that due to the original induction. This is a special case of the "compensating effect" which nearby earth conductors can produce. At low frequencies it is necessary to earth the screen at both ends to provide a complete circuit for the induced currents. Then the reduction is determined by the ratio of the screen resistance and the impedance of the screen earth loop. Generally this reduction is of the order of 15-40% for most frequently used cables which do not contain specially low-resistance or high-permeability layers, when installed in environments like substations or power stations. The reduction can vary significantly at higher frequencies as the transfer impedance of the cable screen and the impedance of the screen/earth circuit change (Klewe, 1958; Jones, 1953; Fowler, 1970; Harrison, 1965). At high frequencies the second earth connection to the screen can become irrelevant. Problems arising from earthing the screen of a multicore cable at both ends will be affected by whether or not the screen is used as a return conductor in any of the circuits.

Although by making two earth connections to the cable screen it is possible to reduce the common mode voltages in the cable, there are disadvantages in introducing the second earth. It exposes the cable to the possibility of damage by providing a path for stray earth fault currents. This may be particularly a problem in situations such as transmission substations or power stations where there may be high concentrations of earth fault currents, and the cables are therefore commonly constructed such that they are able to carry the appropriate proportion of fault currents. Although it is possible to construct cables in this way with screens and armouring with an adequate cross-sectional area to withstand the expected fault currents, this may not always be the most effective way of countering the problem of common mode interference. The second earth connection also means that there will be an additional voltage difference between the cores and the screen determined by the screen current and screen impedance. This can increase the level of series mode interference if there is unbalance between the cores in the cable. As a general rule it may be the best policy in the first instance to use instrumentation whose circuitry is balanced with respect to earth and which can operate successfully with a common mode voltage on its input connections. In some circumstances this may not be an accpetable solution because the level of common mode voltage is too high, either for personnel safety or for damage to the insulation of the instrumentation. Then it will be necessary to take steps to reduce the effect of the common mode voltage. Before double earthing of cable screens is decided upon, it would be preferable to investigate the alternative methods such as optical links (see Sections 4.5.4.3, 4.5.9) or isolating transformers and devices referred to in Section 4.5.

Where earth potential differences can result in excessive cable sheath currents at power frequency, a method described is to earth the sheath at the protective relay end and to earth the insulated cable sheath through a cpacitor at the switchgear end.

The increasing importance of system engineering aspects has been recognised in the reduction of interference in nucelar plants (Cox, 1975). In this case measuring systems may have to detect continuous currents of the order of 10^{-12} A or pulses less than 100 ns long and of a few hundred microvolts amplitude. Imperfections in the screening are often found on installation; and when cable lengths exceed about 50 feet the screening properties of the cable itself can be a limiting factor, owing to the flow of earth currents in the cable screen (Harrison, 1965). Such currents can be up to 100 mA at any frequency up to at least 100 MHz.

The screening property of a coaxial cable depends upon its surface transfer impedance, which is defined as the voltage between the sheath and center conductor for one ampere of current flowing in the sheath per meter length. Mineral-insulated solid copper sheathed cables have a transfer impedance of about 20 mΩ at low frequencies. Above 500 kHz the figure improves because of skin effects to less than 1 μΩ above 20 MHz. A conventional braided sheath polythene cable cable has similar properties at low frequencies but has worse performance above a few MHz. It is important to remember that MICC cable cores are parallel whereas in other cables the cores are "laid" or twisted. This results in an inferior performance of MICC cable to low-frequency magnetic interference (see MOD electromagnetic compatibility guide [MOD, 1968]). "Superscreened" cables (Fowler, 1970) are now widely used, and these employ one or two mumetal tapes between two or three copper braid sheaths. The transfer impedance is about 7mΩ at 100 Hz and falls rapidly to below 0.1μΩ at all higher frequencies. The cable is more than 10^5 times better than a single braided cable at frequencies above about 100 kHz. Harrison (1965) has shown that instrument cables should have a low value of screen transfer impedance, necessitating low-resistivity material for the screen, maximum screen thickness and high screen permeability.

Apart from the above references, further work on interference in the nuclear and other fields has been published, and many of the findings concern the high-voltage-substation case (see for example: Harrison, Wilson and Cox, 1965; Fowler, 1974; Jones, 1953; Campling, 1977; Vogl, 1977; Seljescth II., 1977; Malmstrom, 1977). Information on test methods of the effectiveness of the screening of cables is given in Chapter 10.

4.2 Electric Fields

Owing to the presence of unscreened high-voltage conductors comprising lines and busbars, the conventional substation contains strong electric fields. Capacitive couplings thus exist to signal conductors in the vicinity, and these must therefore be screened. The screening enclosures already considered in Sections 4.1.1 and 4.1.2 in respect of electromagnetic radiation serve to provide an earthed electrostatic screen within which the electrostatic fields will be reduced. However, because of the large size of the system, and particularly the extent of the cable cores, induced interference within the cores can result (see Section 4.5.1) (Harrison, 1965).

The main recommendation for avoidance of problems from electric fields is to ensure that the whole system is enclosed in a metallic screen. In a cable system, because of the difficulties of insulating this from earth at all except one point, other means have to be sought of avoiding interference with the signalling system itself. In practical cases, stray capacitances often provide some kind of multiple earthing except in the few instances where the terminal devices are physically small.

Where electronic systems themselves are concerned, they will *preferably be totally screened.* If this is not possible, and the design permits, it is worth filling all unused areas on printed circuit cards with a conducting "groundplane." Similarly, sensitive circuits in, for example, an edge connector can be guarded by earthed contacts and conductors placed on either side in neighboring positions. These should be insulated from local earth and returned to earth by a single conductor. In card construction schemes it is often possible to sandwich sensitive parts of the circuit between screening plates or printed circuit cards used only as a screen or counterpoise. *Such screens must be insulated from local earth.* Where common power supply lines at the same potential are employed, they should be run separately and insulated throughout with only a single earth connection point. Common power supply return circuits which use the chassis or metalwork as a conductor should be avoided wherever possible. Similarly signals should not be communicated through screening of cables even if they are nominally at the same potential. This rule is often broken on VHF and UHF radio systems where coaxial R.F. connections are common. Under substation conditions this can introduce noise due to the flow of noise currents in the cable screen (Fig. 18 and Section 4.5 refer). Twin balanced cables are preferable for this duty.

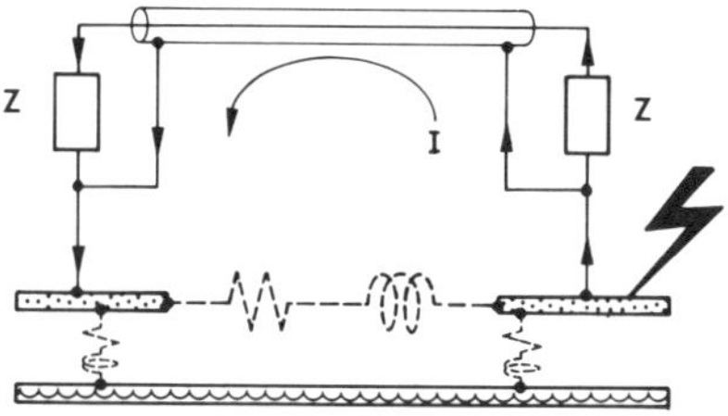

FIG. 18 CABLE BALANCE – CIRCULATING CURRENTS CORRUPT DATA

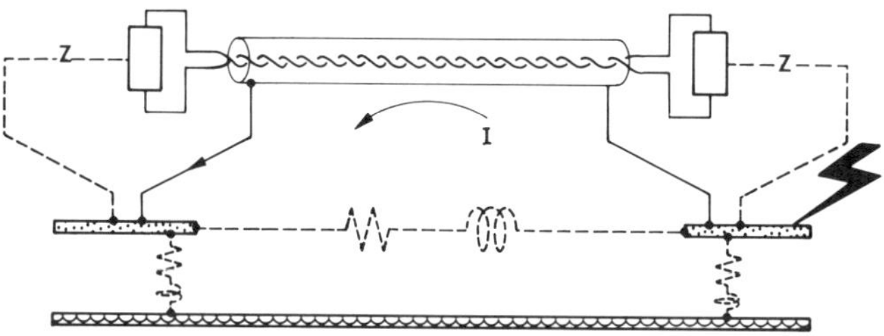

FIG. 19 CABLE BALANCE – IF IMPEDANCES Z AND CABLE ARE BALANCED WITH RESPECT TO EARTH DIFFERENTIAL MODE INTERFERENCE IS REDUCED. SCREEN IMPROVES COMMON MODE REJECTION

4.3 Problems with peripheral cables

In substations, high transient rises of local earth potential are common particularly during switching operations; high currents with short risetimes giving a spectrum extending to several tens of MHz occur in the high-voltage conductors and also in the earth mesh, which includes the many cable sheaths some of which enclose the low-level signalling circuits to the peripheral equipment.

Taking a simplified view, there are four causes of interference with electrical signal cables. A detailed explanation is given in the literature (Anders and Campling, 1976). Electrostatic pickup (considered in Section 4.2) is solved relatively easily by adequate screening. Inductive pickup can be reduced by employing twin twisted pairs in the signal cables. Pickup due to earth loop currents in the screens is reduced by employment of balanced signal transmission. Lastly, common mode interference, which may result from conductive coupling (rise of earth potential) or inductive coupling, can be converted to an undesirable differential mode by cable leakage (Harvey, 1970) or by nonsimultaneous breakdown of protection devices. This may be countered by appropriate barrier devices (Fig. 3).

4.3.1 Balanced transmission

Preferably, signal transmission should be arranged to be electrically balanced about earth over the whole frequency range under consideration. This is difficult and often impossible to achieve. It necessitates a close control of symmetry in the cable pair and in both the source and sinking systems attached at each end (Figs. 18, 19). At all interference or signal frequencies above 10 kHz, it is now easier to improve screening (by the employment of superscreened cables) than to incorporate balancing transformers or to change intrinsically unbalanced circuits.

Inductive pickup results in a differential noise voltage, and like electrostatic pickup it is proportional to the length of the cable exposure and roughly inversely proportional to the distance between the heavy-current "primary" and the signal cable under consideration. Inductive pickup is minimised by ensuring maximum spacing, laying signal cables at 90° to the heavy power conductors and above all by tightly twisting each individual pair throughout the length.

The mechanism of common mode interference is explained in Section 4.1.2 and in the literature (Harvey, 1970; Bull, 1971, 1975; and documents listed in Chapter 2). Where balanced signal transmission cannot be employed because of limitations in existing circuit design, then there will be a risk of interference ingress (McDonald, 1966; USAF, 1965; MOD, 1968).

4.4 Choice of Signalling Methods

4.4.1 Signal levels

In a substation environment it is clearly essential to use the highest signalling levels. For example, transmission of millivolt level analog signals, say from a temperature measuring thermocouple, over a long peripheral cable into an electronic logging system would be undesirable in the substation environment. It would be preferable to condition such signals to a higher voltage or current level by a suitable device located at the thermocouple end, and to transmit high-level signals over the cable thus maintaining an improved signal-to-interference ratio over the part of the route having the greatest exposure.

4.4.2 Filtering

Similarly, D.C. circuits, for circuit-breaker position indication for example, provide a broadband path for the ingress of both common mode and perhaps differential mode interference. One solution is to employ appropriate barriers and filters.

4.4.3 Use of carrier frequencies

Another possible solution is to utilise relatively high carrier frequencies for such indications, thus confining the signals to a part of the spectrum less prone to interference from the substation environment and facilitating the rejection of noise by the incorporation of sharply tuned filters accepting only the one carrier frequency. An advantage of this approach is that common mode barrier R.F. transformers are easily and cheaply made. They permit minimum capacitance together with the incorporation of electrostatic screening. This scheme may usefully be applied in the transmission of switch position indications where the traditional D.C. methods may otherwise necessitate the use of barrier systems using relays or optoisolators, and are useful where the interference energy is distributed over a wide spectrum, thus taking advantage of the narrow signal filtering to exclude the major part of the interference energy.

4.4.4 Optical links

An attractive way of avoiding induction and potential rise problems is to entirely replace the cables to the peripheral equipment with optical free space or fibre links (see Section 4.5.9) (Rogers, 1975). These are of course constructed from a dielectric material and are thus largely unaffected by the problems considered in this book (see Entries 9, 31, Appendix A).

4.4.5 Signalling within a vulnerable system

The problems arising from the use of unbalanced signal transmission are not confined to the peripheral cables alone. Many computer manufacturers have used unbalanced signal transmission within their computers, and this can lead to corruption of data due to external conducted interference (Section 4.5.2).

4.4.6 Signal processing

It will frequently be found possible to tolerate worse signal-to-interference ratios than at first sight seem acceptable by the inclusion of appropriate filtering and signal processing. For example, where time permits, integration within an analog-to-digital convertor can achieve a marked improvement of power-

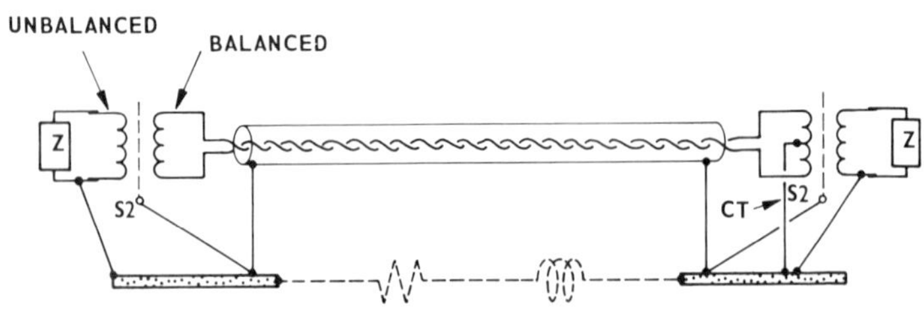

FIG. 20 BALANCED TWIN TWISTED A.C. SIGNAL CABLE

frequency noise rejection (Harvey, 1970; Wheable and Outram, 1970). Similarly, where computer processing of the information is used, a number of additional techniques can be employed to extract meaningful information from a noisy signal.

4.5 Signal Circuits—Rise of Earth Potential, Common and Series Mode Interference

As local transient rises of earth potential in a substation can reach values of several kilovolts, currents will enter galvanically connected cable circuits and will also be coupled capacitively and inductively from cable sheaths in the affected area into the cable cores. Such cores may end outside the immediate locality and may thus impress high voltages on any connected equipment.

4.5.1 Transmission modes in cables

If unbalanced signal transmission is used, then a prevalent source of difficulty can arise due to circulating currents in the signal path (Fig. 18). Providing that the source and sink impedances can be maintained balanced about their local earths over the whole frequency range of interest, the arrangement of Fig. 19 is preferred (see also Section 4.1.2).

Because many sources are not balanced, and where A.C. signals are involved, the method of Fig. 20 is frequently employed. The balun transformers provide the requisite impedance matching and ensure that conditions on the cable are maintained balanced and truly "differential about earth." Usually it is desirable to provide a single galvanic earth to tie the cable pair electrostatically to the ground potential. This connection, however, plays no part in the signal communication itself. However, the quality of the balance of the source and the sink has to be maintained over the full frequency band, as otherwise frequency-sensitive imperfections in the balance can lead to enhanced interference.

4.5.2 Transmission modes within vulnerable equipment

The undesirable effects of using unbalanced signal

transmission are not confined to the peripheral cables; they can occur within the cabinet containing vulnerable equipment itself. A computer-system comprising two elements A and B is shown diagrammatically in Fig. 21. These two elements communicate with each other over an unbalanced signalling circuit L. Line Y is nominally earthy at each end, being returned to the individual power supply units in A and B and thence to separate earth connections a few metres apart on the enclosing case.

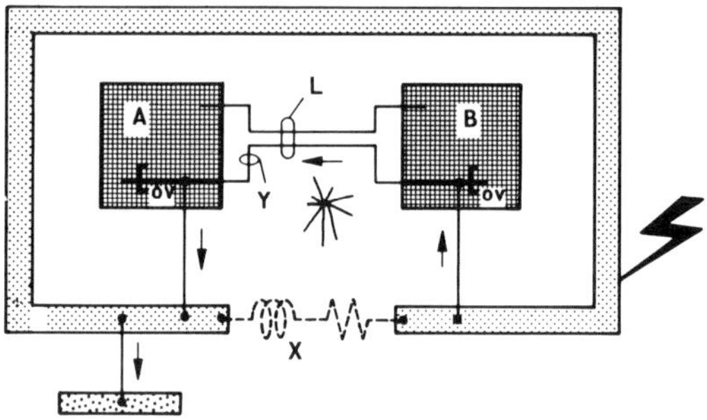

FIG. 21 CORRUPTION OF DATA – EFFECT OF UNBALANCED SIGNAL TRANSMISSION WITHIN AN ELECTRONIC SYSTEM DUE TO MULTIPLE EARTHS

Owing to the finite impedances X, a discharge of static electricity to the case can result in the circulation of a transient current via the line L, which is added to the data being passed. In principle, inductive couplings can also occur.

In an actual case, as balanced signal transmission could not be incorporated, the system was modified (Fig. 22). A clean in-

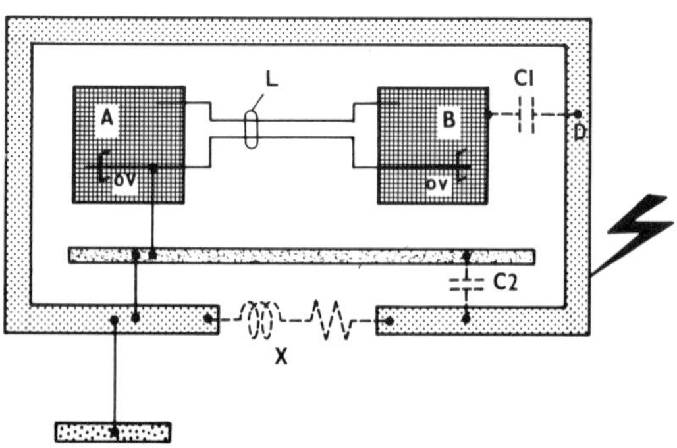

FIG. 22 CORRUPTION OF DATA – AVOIDANCE OF MULTIPLE EARTHS AND USE OF CLEAN INSTRUMENTATION EARTH AS GUARD SCREEN CURES PROBLEM

strumentation earth connected to the enclosure at one point only was incorporated. Although this cures the problem there is still a risk of similar interference entering via the capacitance C1. It is therefore important to extend the clean earth bar as a Faraday screen to wrap round, insofar as is possible, both systems A and B. In electronic systems built with printed circuit cards, this is often possible by using all spare areas of P.C.B.s as a screen or ground plane, arranging that this metal shall be insulated from local earth and returned to the clean earth bar by single connections (see also Section 4.2). Regrettably, some manufacturers use the undesirable practice of returning the 0-volt line through the metal frame. This should be avoided (see also Section 4.5.7).

Multiple earthing of the circuit itself is forbidden, as it represents a loop. Similarly, although multiple earthing of a cable screen may sometimes be useful, multiple earths applied at separated points to an equipment cabinet or enclosure can give rise to difficulties.

4.5.3 Balun transformer screens

Typical values of interwinding capacitance in small transformers may be reduced from 300-1,000 pF to 20-30 pF by the introduction of a single screen. Because of the finite inductance in the transformer interwinding screen connections (Section 4.5.1) and the capacitance of the windings to the screen it is often desirable to incorporate three screens (Fig. 25) in each transformer. The additional screens, which are all insulated from one another, intercept the interwinding capacitance and leakage currents and return them whence they originate.

It should be noted that such screens must be constructed from metal foil. The practice of making a screen by means of a single-ended fine wire single layer winding is often used in cheap transformers. While such a screen can intercept interwinding D.C. leakage currents flowing through the insulation, the residual distributed inductance renders it unsuitable for the application under consideration.

Where A.C. or pulse waveform signals cannot be used, then the isolating-balun-transformer concept is no longer applicable and alternative forms of isolator will be needed.

4.5.4 Barrier isolation devices

These may comprise any of a vast variety of devices such as electromagnetic relays (where operating speed, life and contact bounce considerations permit), isolation amplifiers, optocouplers, radio or carrier frequency links, optical fibre links, etc. In all cases such devices, selected from the large range commercially available, must be considered for their ability to act as a barrier to both series and common mode interference. If the device falls short in the rejection of series mode noise, then additional steps such as filtering by hardware (or software) will be required. The performance in respect of common mode rejection of many barrier devices could be improved by the addition of even a single screen between the input and output sides.

Clearly the highest practicable level (Section 4.4.1) should be used for the signal transmission through long cables themselves vulnerable to interference; and an attenuator or divider, which may often be frequency-sensitive, will be used close to the computer or vulnerable equipment. Typically, resistive potential-dropping networks followed by shunt Zener diodes (Entries 27, 42, 46, 48, 52, 70, Appendix A) or zinc-oxide voltage-dependent resistors (Entry 46, Appendix A) are used to control the voltages applied to the computer inputs and establish them within a range of a few volts only above local reference earth and similarly between lines where balanced signals are employed. The limiting network is usefully preceded by a protector which can be of the nonlinear resistor type, a spark gap or more usually a gas-filled protector. A typical modern gas-filled protector has the following properties:

D.C. sparkover voltage, 230 V
Impulse sparkover voltage at 2kV/μs, 600 V
Impulse discharge current, 8/20 μs form, 25 kA
A.C. discharge current, 1 second 50 Hz 20 A
Dimensions, 8 mm long × 8 mm diameter

Protectors of this type (UC230[Q]) have been employed in

the circuit of Fig. 23 for the protection of logic signals entering integrated circuits in meteorological-data-acquisition equipment (Entry 13, Appendix A). The filter itself provides an attenuation frequency characteristic as shown in Fig. 24. Solid-state transient spike suppressors ("transtectors") can typically be used alone or in parallel with a conventional silicon controlled rectifier crowbar and will give 5 ns protection. A certain maker offers two types giving effectively instantaneous clamping at 6.5 V D.C. 1 A or at 16.5 V D.C. 400 mA. The corresponding peak pulse currents are rated at 120 A and 67 A, respectively (Entry 18, Appendix A). See also Section 4.5.5 for radiofrequency filters. Encapsulated shunt regulator and crowbar units are available for protection of supplies or printed circuit cards from several manufacturers (Entry 53, Appendix A).

Where analog signals have to be transmitted with accuracies of up to ±0.01%, isolation amplifiers offering up to 5 kV isolation in the presence of large common mode interfering signals are available (Entries 5, 12, 16, 19, 33, 44, Appendix A). A 7.5 kV isolation amplifier is also available. Where milliampere type analog signals are involved, isolation devices giving 1 kV or more isolation are available. One needs no local supply (Entry 54, Appendix A).

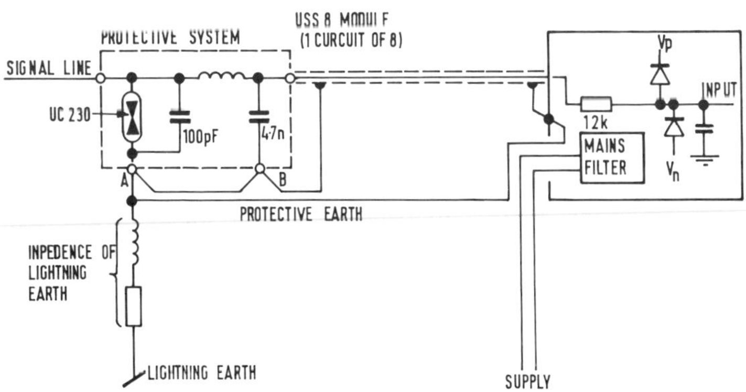

FIG. 23 EXAMPLE OF PROTECTION DEVICE USING GAS FILLED PROTECTOR (after Cerberus leaflet N°38 Sept 1975)

The application of various barrier devices is summarised in Table 1:

Table 1: Application of Barrier and Protection Devices

Device	Effective against interference: Common Mode		Series Mode: Low Level	Series Mode: High Level
Relay		Yes*	Yes	Can falsely operate with interference
Transformers		Yes*	No	No
Optoisolators		Yes*	Yes	Can falsely operate with interference
R.F. Filters		Yes, if of appropriate type	Yes	Yes
Optical-Fibre Links		Yes	Yes	Yes
Isolation Amplifiers		Yes	No	No
Limiters (Diodes, crowbars, discharge tubes)	Each line to earth	Yes, but unless simultaneous conduction is assured, series mode will be produced by the resulting unbalance	No	Some protection
	Line to line	No	Yes	Yes

*Faraday screen(s) are desirable

It is frequently found that the small physical size of certain commercial devices makes it difficult to achieve adequate segregation between the wiring associated with the input and output sides. This comment relates not only to the physical separation necessary to avoid electrical breakdown but also to the reduction of mutual inductive and capacitive couplings. It is not helpful if such devices are installed so that the input and output wiring is combined into common connectors or cable forms. By making the screen between the input and output of a barrier device effectively a part of the vulnerable system screening enclosure, the ingress of electromagnetic interference can also be prevented. Unless this is done it may often be necessary

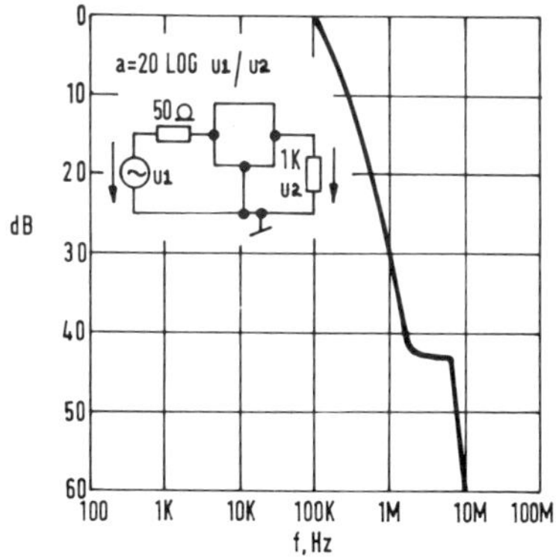

ATTENUATION-FREQUENCY CHARACTER OF CERBERUS METEOLABOR USS 8 MODULE

FIG.24 FILTER ATTENUATION CHARACTERISTIC

to insert additional radiofrequency filters at the point where cables enter the enclosure (Figs. 8, 9).

Many special-purpose barrier devices are available for protection of specific items such as modems and communication lines (Entry 48, Appendix A). Where telecommunication lines enter an area where rise of local earth potential is likely to be encountered, a practice common in some countries involves the use of neutralising transformers. In one type the two incoming conductors of the telecommunication line are connected to the station equipment via two windings wound in the same sense. A third winding in the opposite sense is used to inject an opposing voltage, and it joins the station earth to a remote earth. Such devices, which are considered in detail in an IEEE Interim Committee Report F76 379-8 (IEEE, "The Neutralising Transformer Concept...," 1976) can clearly only be useful at power and audio frequencies. The report also gives several references on this topic.

4.5.4.1 Relays

Historically, isolation of binary signals has often been achieved by the use of conventional or reed relays (Entry 1, Appendix A). Where speed is unimportant and where arrangements can be made to eliminate the effect of contact bounce (Motorola MC 14490 device) relays can provide excellent isolation, subject only to life and thermal limitations. However, in many commercial forms electrostatic screening between the input and output sides is absent or poor. Furthermore, where low-level signals are being switched, induction may occur into the contact circuit. Input/output isolation levels of 10^{10} ohms, 1.5 pF and 1,000 volt A.C. have been claimed for mercury wetted relays (Entry 15, Appendix A) (Clare, January 1975).

The wiring on the two sides should be well separated (not in a single cable loom or even in a single enclosure) and where possible should be arranged at 90° one to the other and, if possible, segregated by a ground plane. Where balanced circuits are employed, open loops should be avoided by keeping the two twisted wires of the input pair close together, the two wires of the output pair being also close together and well separated from the former.

A method of providing the desired electrostatic screen(s) between the input and output is to use a multilayer printed circuit card with the input and output circuits each on one of the two surfaces between which are sandwiched the segregated electrostatic screens. Magnetic screens may also be necessary to prevent crosstalk between closely packed adjacent reed-relay elements. Ideally, three screens are beneficial. Where there is freedom of choice in the form of signal (Section 4.4), the improved reliability of transformer barriers working at a carrier frequency encourages their use rather than relays which have limited life (Section 4.4.3). More recently, various types of solid-state relay have appeared and are claimed to have a high noise (interference) immunity (Entry 67, Appendix A).

Galvanic isolation is now included in many commercial systems such as the Indactic system (Brown Boveri, 1974) and the CAMAC system (Borer) (Entry 11, Appendix A), and is clearly an essential precaution when electronic equipment is interfaced to peripheral equipment in an electrically noisy environment.

The C.E.G.B. have specified a reed-relay interface having a 10 kV isolation for use in protection signalling/intertripping applications (C.E.G.B.[a], 1972). The devices do not at present contain an electrostatic screen, which is a possible disadvantage as high-frequency interference can easily pass a few picofarads capacitance. The reed relays themselves are specified in a C.E.G.B. document, "General Specification for Reed Relays" (C.E.G.B.[b], 1974) (see also Design Guide No. 4 Issue 3 [C.E.G.B.(c), 1973]). A modular component holder ECHI has been produced to meet C.E.G.B. requirements, and this provides a convenient housing for limiting, isolation, attenuating and filtering components. The holder is designed for assembly in a modular terminal block housing. However, in this design the lack of enclosing screening can result in capacitive couplings to neighbouring elements. A variety of devices are offered by manufacturers (Entries 7, 21, 35, 38, Appendix A). Where A.C. signals have to be switched (up to 40 amperes and 600 volts peak blocking capability), optoisolated solid-state relays with high transient immunity are available. Such relays typically provide 1.5 kV A.C. withstand at their output and have a 10^{10} ohm 8 pF barrier (Entry 67, Appendix A).

4.5.4.2 Transformers

Where A.C. signals, such as analog values of voltage and current, speech frequencies, etc., are involved, transformers are an effective barrier if properly designed to minimise capacitive couplings which can degrade the isolation (Entry 73, Appendix A).

Toroidal transformers are preferred because their field is more closely confined and they are thus less susceptible themselves to inductive pickup of spurious signals from the local environment.

Because of the small size of higher-frequency transformers, the use of carrier frequencies is attractive (Section 4.4.3). Isolation transformers should preferably be provided with triple screens connected as shown in Figs. 25 and 26 (see also Section

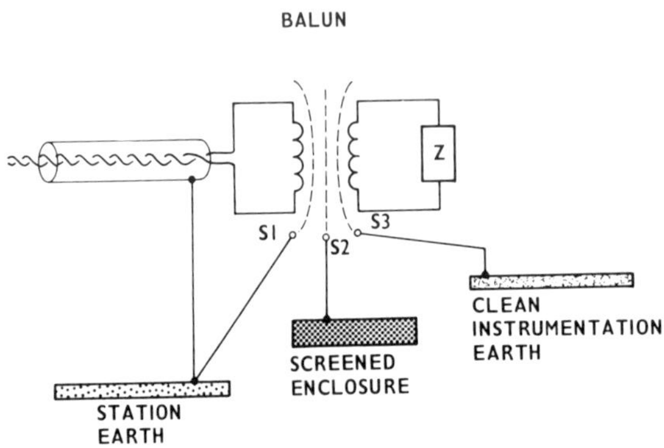

FIG. 25 TRIPLE SCREENED BALUN

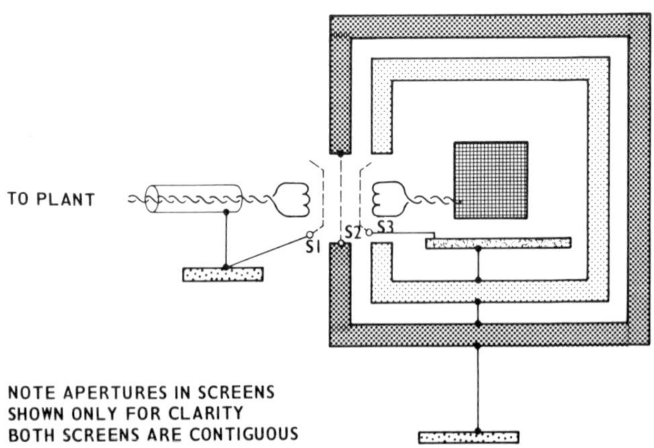

FIG. 26 TRIPLE SCREENED BARRIER TRANSFORMER

4.5.3). Small transformers provide an attractive solution to the blocking of common mode interference in both digital and analog communication paths. Transformers protect against common mode signals only but have no influence on differential mode interference which is passed through with the signal. Attention is drawn to the need for the Faraday interwinding screens in transformers to be substantially noninductive (see Section 4.5.3), segregation of the terminals and wiring and individual earth connections for the screens being provided. Pulse transformers can provide a useful solution in thristor and similar trigger circuits (Entry 69, Appendix A).

4.5.4.3 Optoisolators (Optocouples)

Where binary signals or analog signals that can be encoded in a binary form are to be transmitted, optical isolators provide a useful solution. Many types are available, typically with insulation withstand levels of 2.5 kV and very low transfer capacitances of the order of a few picofarads. Some types can withstand 10 kV. Due to their small size, low cost and absence of moving parts, they are an attractive solution. It is not normally possible to incorporate a segregating electrostatic screen within the device itself. However, the screening between the two sides can still be provided on the mounting card or assembly. There is an extensive literature on optoisolators (Monsanto, Hewlett Packard) (Entries 6, 28, 45, 47, Appendix A). The reported gradual degradation of performance of self-contained optoisolators, the difficulty of inserting an electrostatic screen in most commercial varieties to eliminate the capacitive couplings, and reported problems of false operation due to internal leakage to the photosensor give some cause for concern. Unless these problems are solved, optoisolators may remain less attractive for long-life high-reliability uses.

4.5.5 Radiofrequency filters, common and series mode

A wide range of radiofrequency filters is commercially available, and these can be used to give a high rejection of interference at frequencies extending to tens of MHz. Simple through-bushing type radio filter capacitors (Entries 4, 22, 24, 58, 59, 60, 61, 71, 72, Appendix A) can give useful R.F. attenuation by means of the "shunt path to local earth" method. The importance has been emphasised of the choice of a capacitor having low residual self-inductance (Bull, 1975). Where capacitors are used, the effect of this self-inductance has to be taken into account. The self-resonant frequency will depend on the construction of the capacitor and the length of the associated leads. Typical values are shown below:

Capacitance, microfarads	Length of lead to each terminal, inches	Self-resonant frequency, MHz
4	1	0.35
4	4	0.18
0.25	1	1.5
0.25	4	0.7
0.01	1	7.5
0.01	4	3.7
0.0001	1	25
0.0001	4	11

R.F. filters built into screened multiple-contact connectors form a particularly economical element for treatment of leads entering a screened enclosure (Appendix A).

A sound method of mounting through-bushing filter capacitors is to incorporate them in a copper or brass plate forming a part of the screened enclosure (Figs. 8, 9). Methods of mounting and performances of the many types have been illustrated (Siemens, 1969, 1970; MOD, 1968).

If capacitors of any type are used, they should be of substantially noninductive construction specially designed for R.F.I. suppression and above all should be connected to a single common point by low-radiofrequency-impedance connections. All connections must be kept short, and the earth point must be common to all circuits (MOD, 1968).

The loss angle of a capacitor is an important guide to its quality, and a useful discussion of methods of measurement is given by Tait (1978) in the U.K. journal *Electronics Industry.*

Special instruments for measurement of capacitor losses are offered by several manufacturers. (Entry 20, Appendix A).

The radiofrequency voltage in any conductor must be divided between the internal R.F. impedance of the interference source and the interference suppression capacitor which is connected in shunt with the source. When the internal impedance of the interference source is inadequate, an interference suppression inductor must be inserted in the conductor. The R.F. characteristics of the capacitor will depend on its capacitance value, the type of construction and most importantly on the length of the leads. Long leads increase the residual inductance. Feedthrough capacitors become essential for broadband applications, as the impedance of conventional types increases markedly above a few MHz instead of falling with increasing frequency according to the theoretical curve of $\frac{1}{\omega C}$. With suitable types, impedances of a few hundred milliohms can be achieved at frequencies of tens of MHz (Siemens, 1969/70).

The efficacy of a suppression capacitor may thus be characterised by its transfer impedance at various frequencies. Where entry to a screened enclosure is involved, through-bushing capacitors are a common choice and are available with current-carrying capacities up to 100 or more amperes. Such capacitors will usually be used in conjunction with inductors (Siemens, 1969, 1970) to achieve adequate rejection. Alternatively, complete filter units may be obtained with attenuations typically in the range 60-100 dB over the frequency range 100 kHz to 30 or more MHz, with current-carrying capacities up to 100 amperes (Entries 8, 22, 38, 46, 48, 52, 58, 61, 62, 67, 68, 72, Appendix A). The fundamental aspects of achieving appropriate attenuation in radio noise suppression have been considered in a 1966 paper (Mayer and Faure, 1966), and although aimed principally at vehicle suppression that paper contains valuable guidance. Considered together with that of Harrison (Harrison, 1965), it is a useful adjunct to the ERA report 75-31 (Bull, 1975), and MOD references.

The higher attenuations can only be achieved by using several sections in cascade, and these invariably need to be individually enclosed in a multicompartment box contiguous with the screening enclosure. It is recommended that toroidal inductors should be used in preference to solenoidal windings so as to reduce extraneous fields and susceptibility to external fields.

Where increased impedance at high frequencies is a requirement, ferrite ring cores fitted over the conductors provide a useful solution. It should be noted that ferrites used in this way can only be effective if there is a low-impedance shunt path. This has been applied by Kingham (James, Kingham and Moore, 1973) in reducing lightning-discharge currents flowing in sheaths of instrumentation cables on a 385 m high radio mast, and "braid-breakers" are also a common practice in the amateur-radio field (Jackson, 1972).

4.5.6 Location of filters and barrier devices

Many filters and barrier devices are not used to the full owing to incorrect installation. Fig. 27 indicates an appropriate method. Preferably, the filter unit will be mounted on the screened enclosure and close to the common earth point E. If this is not possible, the filter must be mounted as close as possible. The connections on the vulnerable equipment side must be screened. If the wiring on the two sides of the filter are brought together and unless at least one side is screened, then the attenuation may not be achieved. It is emphasised that the gaps in the ends of the screening of the cable shown in Fig. 27 are shown purely to clarify the diagram. In practice, a connection to earth of the type illustrated is not admissible. Shell-type connections by which the cable screen is maintained radiofrequency-tight at the junctions are obligatory. Such a connector is shown in Fig. 28. Attention is drawn to Section 4.2 of the ERA report 75-31 (Bull, 1975), which indicates suggested minimum values of attenuation for mains-borne noise in the general, as opposed to the hostile substation, environment. Filter attenuations of 60-100 dB are readily obtainable.

Fowler considers that, for a connector to be acceptable, not more than 10% of the interference, when injecting a test signal of 100 mA rms in the cable screen, should arise at the connector due to penetration through the transfer impedance, and that no more than 5% should arise due to the mating face (Fowler,

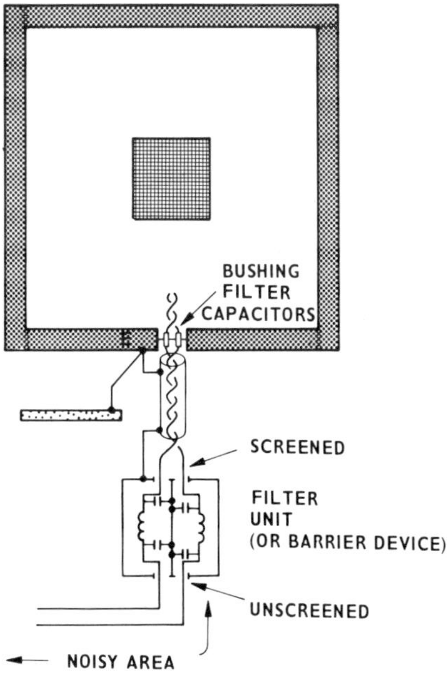

FIG. 27 INSTALLATION OF FILTERS & BARRIER DEVICES

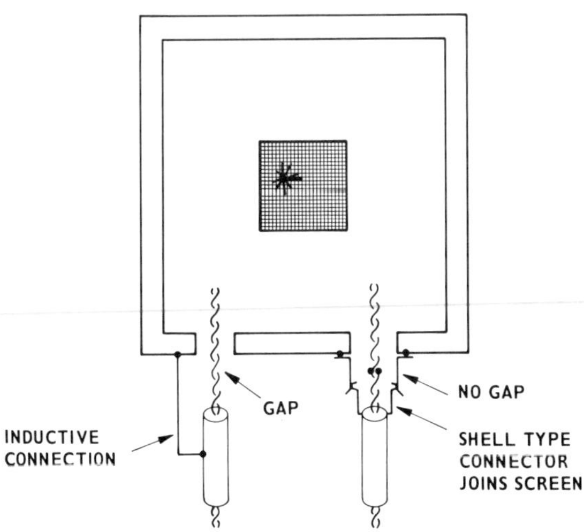

FIG. 28 BONDING TO ENCLOSURE

1974). The correct choice of connector and its mounting on the enclosure present considerable difficulties with low-level broadband circuits where connectors having mating-face impedances of less than 5 micro ohms may be a requirement (see also Chapter 10).

4.5.7 Electronic equipment design, chassis and return circuit

Traditionally, many electronic systems have used the chassis as a return circuit, particularly for the power supplies or 0-volt line. This is acceptable in domestic equipment and in many commercial installations but is undesirable in certain forms of equipment in the substation environment. Stray earth currents connected either galvanically or induced into the chassis can cause malfunctions. Furthermore, impedances presented by the chassis can give rise to galvanically connected stray currents, which, because of the multiplicity of separated connections to the chassis within the electronic circuit, can flow into the circuit itself, as well as giving rise to loops vulnerable to induced circulating currents.

If possible, the electronic system should always be designed to contain its own clean instrumentation earth which is joined to the chassis and the enclosure at one known point only, no reliance being placed on the metal chassis or frame except as a mechanical support and electrostatic screen. However, this instrumentation earth is not a separate earth electrode and is bonded at one point to the general earthing system within the apparatus room. Many manufacturers could raise the interference immunity of their products by attention to these points.

Because of the undesirable practice of using unbalanced signal transmission between various units, metallic signal connections may exist in parallel with power-supply connections feeding the individual units from a common power unit. Similarly, if the power-supply line and a common signal line loop into each unit (in series), series mode volt drops can give rise to difficulties. The solutions are to employ one of the following alternatives:

(a) Use separate power supplies.
(b) Use balanced signal transmission between each unit.
(c) Incorporate series mode barriers in the signal transmission lines (e.g., optocouplers, transformers, etc.)

In all cases the power supply 0-volt line(s) will be independently returned to a single clean earth point (see also Section 4.5.2).

It is also good practice to provide individual decoupling elements on each circuit card where power supplies are taken from a common rail. By this means the conduction of transients into the electronics by way of the common supply is minimised (Entry 53, Appendix A).

Similarly, where choice exists, local power units incorporating appropriate barriers on their input sides are preferred to the distribution of several smoothed supply rails from a common source.

Where low-level circuits within an extensive electronic system are employed, the circuit should where possible be balanced; and even if this is not possible the two sides should always be maintained close together (preferably twin twisted) to minimise magnetic induction into open loops. Circuits should never pass through different go and return cables but should be always within one cable (see Section 4.1.2).

4.5.8 Mounting of electronic equipment, earth isolation

Where the electronic system is already built with multiple internal chassis connections, then external interfering stray earth currents through the chassis can be minimised by insulating from the supporting rack and bonding it *at one point only.* In many types of proprietary computer, for example, the 0-volt line may have multiple connections to the chassis; so steps have to be taken to incorporate galvanic insulation of the complete electronic unit itself, which should be insulated from the supporting rack so that currents flowing in the racking are not able to enter and leave the electronic chassis (except by way of the small residual capacitances and inductive couplings) (Fig. 21). A single earth connection should be made, as in Fig. 22, to the 0-volt line on the computer.

4.5.9 Optical links

Apart from the electromagnetic screening of the vulnerable equipment itself and the insertion of barriers in its power supply, most of the considerations in this book arise because of the assumption that *electrical* signals will be used to link the peripheral systems to it. If a method of communication itself unaffected by interference could be employed, all the precautions (and the unpredictability in practical situations) could be dispensed with.

Such a method, which is now completely practicable (Andersson and Anders, 1977; Sandbank, 1973; Brace and Ravencroft, 1977; Midwinter, 1976), relies on the use of electrically insulating optical-fibre bundles to convey light signals. Rogers (1975) has given a detailed survey of this technique, and it must be mentioned here as a practical solution to all forms of signalling within noisy environments.

Optical-fibre-bundle cables and free-space optical links are commercially available, and the transmitters and receivers interfacing the system to conventional electrical equipments at distances of up to hundreds of meters are simple and cheap (Entries 3, 9, 26, 30, 31, 32, 34, 44, 49, 55, 63, 66, Appendix A). Nevertheless, it should be noted that the receiver operates at low electrical levels, and it will require screening as well as appropriate filtering and compact electrically balanced design to reduce conventional interference to an acceptable level.

This method is undoubtedly gaining ground as an elegant solution. Although electrical signals have been chiefly used in the past, this does not mean that the optical method may not largely supplant current methods in the future.

4.6 Power Supply and Instrument Transformer

Because the vulnerable system of necessity requires an electrical power supply, any external connection of this supply is a potential route for both series and common mode interference ingress unless isolation is provided. Similar precautions are applicable to instrument transformer circuits.

4.6.1 Self-contained power sources

Clearly, if the power source can be totally self-contained within the screened enclosure of the vulnerable equipment, as by the use of replaceable batteries, the problem is avoided. However, this is rarely practicable, although equipments in which the power source is brought in from externally in a nonelectrical form have been utilised. Examples powered by a compressed air or oil pressure powered turbine driving a small generator within the enclosure exist, the power being supplied through an insulating pipe. Similarly, solar-cell recharging of a secondary battery has been used. Other alternatives are to use propane-gas-fired thermoelectric or thermomechanical converters giving an electrical power output. (Cooke-Yarborough, *et al,* 1974a, 1974b). Small systems of this type offer the advantage of confining the effects of any electrical fault to the equipment concerned rather than impressing the resulting transients on all other equipments joined to a common power-supply rail. Furthermore, the thermal time constants result in a high attenuation of series mode electrical transients in the supply voltage. Common mode transients are of course totally eliminated by the insulating gas pipe.

4.6.2 External power sources

In practical cases, most equipments will at present be powered from an external source of supply. The only course open is to insert adequate common and series mode barriers/filters to ensure that difficulties cannot occur. Hooper has analysed a particular case of a 240 volt A.C. supply brought to a 400 kV line tower (Hooper, 1973), and he makes specific recommendations capable of dealing with the effects of large rises of earth potential and also of lightning.

In the case of supplies taken from a 240 volt transformer, for example, interference ingress due to rise of earth potential is a serious problem, due to the remote neutral earth point, unless appropriate precautions are taken. Fig. 29 indicates a typical

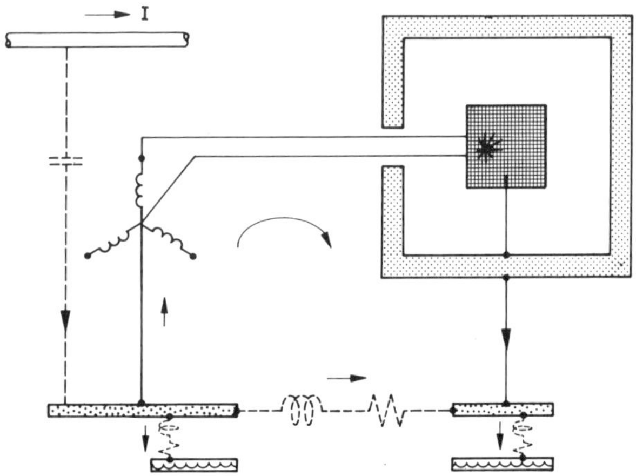

FIG. 29 POWER SUPPLY WITH NO PRECAUTIONS

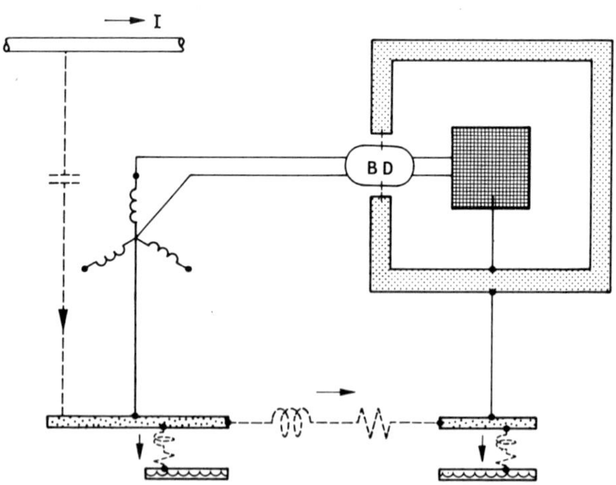

FIG. 30 POWER SUPPLY WITH COMMON MODE BARRIER

scheme in which a remote transformer neutral earth forms a nonconductive earth loop via the impedances between the mains wiring and local earth within the vulnerable equipment. In Fig. 30, a common mode barrier is inserted to prevent circulation of earth currents.

Such barriers are readily constructed exactly as for A.C. signal circuits. A single screened transformer may be used, or preferably three screens connected as shown in Fig. 26.

Where preferred, two single screened transformers connected in cascade may be employed (Fig. 31). In all cases, toroidal transformers are preferred owing to the fact that the field is more closely confined. Owing to the practical problems of constructing triple screens on toroidal transformers, the two transformer cascade connection is then preferred.

Precautions may also need to be taken against series mode spikes due to induction from power-system circuits and to the switching of other loads. Usually, the electronic stabilisers within the electronic equipment are adequate for elimination of problems caused by the direct effect of series mode power fluctuations. However, the radiofrequency components may be reradiated from the wiring of the power unit *within* the screened enclosure unless adequate radio filtering is effected at the entry point of the power circuits.

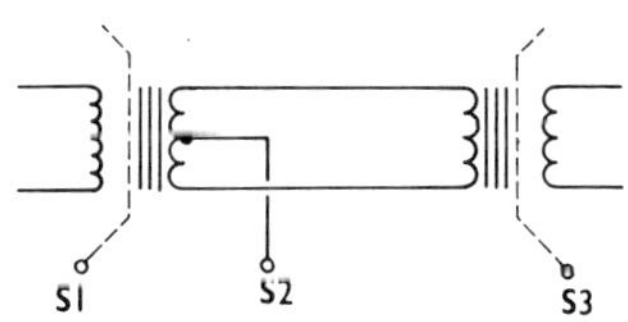

(SEE FIGS. 25 & 26 FOR CONNECTION OF SCREENS)

FIG. 31 TRIPLE SCREEN BY USING CONVENTIONAL TRANSFORMERS

Every power-supply circuit entering the screened enclosure should do so through radiofrequency filters and through bushing filter capacitors fitted through the enclosure wall (Figs. 27 and 30). Where A.C. supplies are concerned, barrier transformers should be used on the supply side of the filter unit.

Such transformers should preferably be triple screened and of toroidal construction (Figs. 25, 26, 31). Capacitors and filters should be of special types with known properties (Entries 9, 51, 58, 62, 68, 71, 72, Appendix A). A typical filter for transient suppression (N.V. Philip MF 2.4 A) gives over 40 dB attenuation between 500 kHz and 10 MHz and being of the balanced type is useful against common mode as well as series mode interference. Relevant specifications are VDE 0560-7 and MIL-STD-202E.

Where D.C. supplies have to enter the screened enclosure, it is usually sufficient to incorporate radiofrequency filters and through bushing capacitors. However, in difficult cases, conversion to A.C. by a suitable external inverter, transmission through isolating barrier transformers, filters and through bushing capacitors followed by rectification within the enclosure has sometimes proved essential (Entry 25, Appendix A).

The radiofrequency impedance of the mains supply is of the order of 50 ohms at frequencies above 100 kHz, and as it is improperly terminated strong resonances may occur at certain frequencies. Fowler (1974) has shown that mains-borne interference can exceed 340 volts peak amplitude; so an immunity to 8 A injected into a properly matched system is appropriate. The mismatch conditions can cause voltages in excess of 1 kV to appear at the mains input terminals. Means of testing the effectiveness of the mains filtering and screening arrangements are referred to in Chapter 10.

4.6.3 Filtering and decoupling within electronic system

Clearly, any commonality within the electronic system which permits propagation of radiofrequency energy through the ramifications of an extensive supply rail (whether earthy or not) is highly undesirable. The more extensive such a conductor, the larger will be its capacitance to the chassis or case; so it will be more susceptible to fluctuations of case potential. For this reason, all common supply rails should, wherever possible, incorporate local radiofrequency decoupling components, individually feeding each card or section of the load. Electronic stabilisers are also cheaply incorporated in each card ("on card regulators") so that common impedances are no longer of significance in coupling transients within the system itself, providing that sufficient response speed is specified.

The main requirements for busbars feeding power supplies within electronic equipment are low inductance and high distributed capacitance (usually between one another rather than to local earth). This can often best be achieved by the use of flat strips positioned and dimensioned so as to achieve low inductance, low resistance and high capacitance.

The desirability of avoiding chassis return circuits in electronic equipment intended for use in substations has already been mentioned in Section 4.5.7.

4.6.4 Connection of screens in interposing current and voltage transformers

Where one or more interposing transformers are included as barrier devices in connections incoming from the main current transformers and voltage transformers in the station, these will usually be connected in the secondary wiring near to the vulnerable equipment. Such transformers should comply with Section 4.5.4.2, and their single (or triple) Faraday interwinding screens will be earthed to the station earth in the locality. If such an assembly of transformers is provided with a common earth connection as shown in Fig. 32, the common impedances presented to the passage of capacitively coupled high-frequency currents can result in voltages being impressed on circuits other than the one affected. Individual earth connections from each transformer should be connected as in Fig. 33 and run back to a single point of common coupling on the station earth bar. The common point will preferably be chosen at a point on the earth strip as near as possible to true earth. This is usually at the point where the common earth strip leaves the apparatus room to join the remainder of the earth mesh.

Where V.T. secondary circuits themselves are concerned, it is preferable for these to provide an output balanced about earth *at the source* feeding a twin twisted cable driving a common mode rejector interposing transformer at the load end.

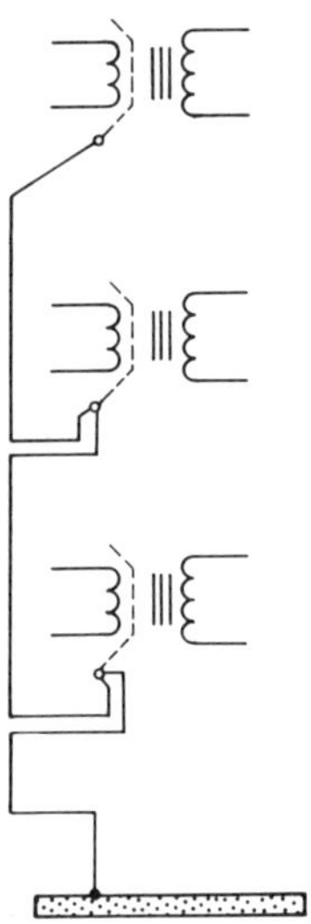

FIG. 32

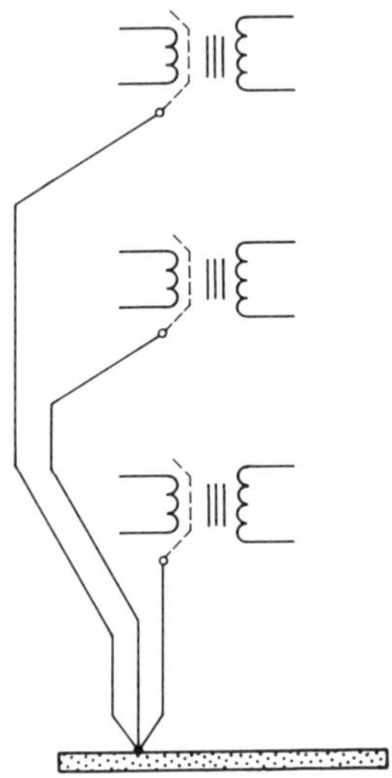

FIG. 33

EARTHING OF INTERPOSING TRANSFORMERS

4.6.5 Connection of instrument earth to station earth

It is clearly undesirable if the instrument earth is tapped onto a point on an earth strip providing a route for transient currents. The earth strip in the apparatus room containing vulnerable electronic apparatus should therefore preferably be arranged as a spur connected to the station earth strip at one point and connected at a point as near to true earth as possible. Fig. 34 illustrates the undesirable case where transient currents flowing through common impedance "Y" from source "C" (and to a lesser extent from sources "A" and "B") give rise to potentials at the clean instrumentation earth. Common impedance "X" is of course inescapable and results from the impedance of the station earth to true earth. However, in practice it is doubtful whether there will be an identifiable point on the station earth mesh at which the inescapable common impedance X is located. Fig. 35 shows the preferred arrangement in which only the inescapable common impedance "X" is still effective. For measurements within the substation, the instrumentation clean earth should preferably be connected at one point only to the station earth as nearly as possible in the centre of the area being covered. In Fig. 35, currents flowing in the station earth from systems A and B will interfere with system C; so for that purpose the arrangement of Fig. 34 would be preferable.

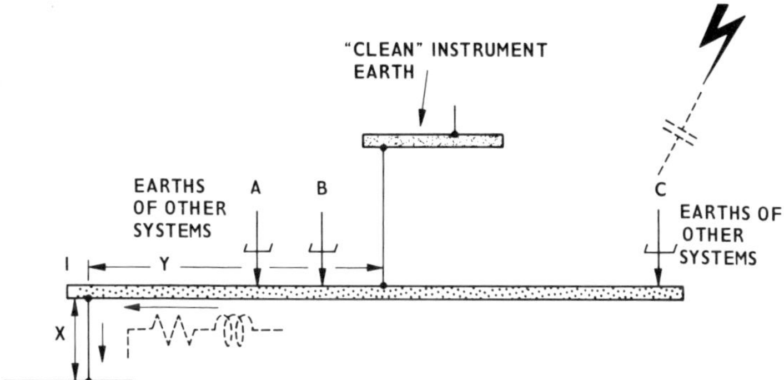

FIG. 34

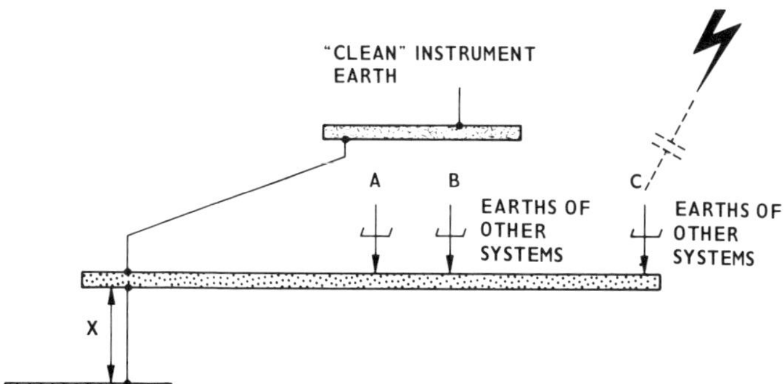

FIG. 35

FIG. 34 & 35 CONNECTION OF INSTRUMENT EARTH TO STATION EARTH

The earth for any instrument is the mass of neighbouring metalwork. It should be noted that attempts to extend the clean instrumentation earth bar to a remote electrode which will be outside the immediate zone of the substation earth mesh and therefore more nearly at true earth potential are unlikely to be helpful. Since the potential gradients are three dimensionally disposed around the station current carrying earth electrode system and the instrumentation area is usually within the precincts, a method of getting a clean electrode which has been tried is that of driving an insulated earth rod down to a sufficient depth in an attempt to take its bare tip beyond the immediate zone of the station earth system, through which the insulated upper part reaches the surface. In practice it is impossible to reach a worthwhile "true earth" in this way, although some users have claimed a limited advantage. In practice we are concerned with high-frequency interference currents, and even a perfectly grounded earth conductor will resonate as a transmission line; so that if we are concerned with frequencies extending to 100 MHz, a length greater than 0.75 m appears to the "earthed" instrument as an open circuit at some component frequencies. Consequently, the best earth system approximates to having a copper sheet to which all equipment is locally bonded. The development of conducting concrete "Marconite" is of interest in this context (Conductive Concrete, 1978).

The measurements made (Fowler and Gregory, 1970) of up to 500 A at 200 kHz in the earth connections near the gas circulator motors of a nuclear generating station illustrate the importance of such measures as the reduction of the inductance of the cable sheath earth.

Because of the large physical dimensions of practical earthing systems in relation to a wavelength at the radio frequencies with which we are concerned, it will usually be impossible to define an equipotential earth point. The best that can be done therefore is to reduce common impedances by taking all earth returns to a single point, so that the station earth system more closely approximates to a continuous earth plane, and relying more on R.F. filtering at the points on ingress to the enclosure. The practical and theoretical aspects of single point earthing and the induced voltages in systems due to the complex impedance of the earth reference line have been examined (Kunkel, 1970a, b).

One of the most useful studies of interference in the reference earth derives expressions for the induced interfering voltages in terms of the parameters of the reference circuit and the interfering source. Mohr (1970) shows a good agreement between his prediction model and experimental data and explains how high-frequency electrically induced interference increases with frequency at 40 dB per decade whereas other types of interference increase at 20 dB per decade. The paper also shows the methods of calculating interference under transient conditions and evaluates the R.F. resistance of certain wire sizes as well as stressing the importance of careful routing and maintenance of close proximity of wires forming a circuit to minimise mutual couplings. Other workers have examined induction by pulsed electromagnetic fields, and their publication evaluates the aspect of multiple reflections in the affected circuit (Naito and Shah, 1974).

Transient suppression in 500 kV installations has been considered (Cushing, 1974). The paper deals in an elementary way with the reasoning for screening, etc., and gives some relative costings of various forms of cable sheathing.

4.6.6 Current transformer secondary earthing

If the current transformer secondary wiring is earthed in the apparatus room as is illustrated in Fig. 36, a highly undesirable situation is produced whereby high-intensity interference is conducted in a common mode against earth di-

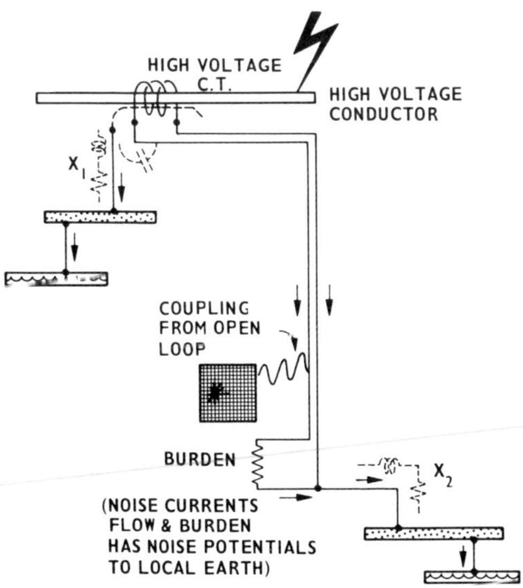

EARTHED IN RELAY ROOM REMOTE FROM NOISE SOURCE

FIG. 36

rectly into the apparatus room. It is well known that the Faraday screens within high-voltage current transformers are only partially effective. This results from the appreciable inductances caused by their physical size and by the inevitable length of the local earth connection from the screen to the station earth. This connection and the screen itself present an impedance "X1" across which capacitively coupled high-frequency currents develop intense interference voltages which are capacitively coupled to the current transformer secondary winding (and possibly to the cable sheath). Such currents return to earth via impedance "X2" of the earth link and its circuit in the apparatus room. In so doing they flow in an open loop, giving rise to interference coupling and to the conducted transverse flow of interference currents in the burden as well as the imposition of voltages to local earth. It has already been shown (Turner, 1968) that current transformers are one of the principal sources of interference voltages. While the introduction of an interposing transformer followed by limiters, suppressors and perhaps by filtering may alleviate the problem, it should be recognised that the entry wiring and the primary side of the interposing transformers still constitute a means of coupling intense in-

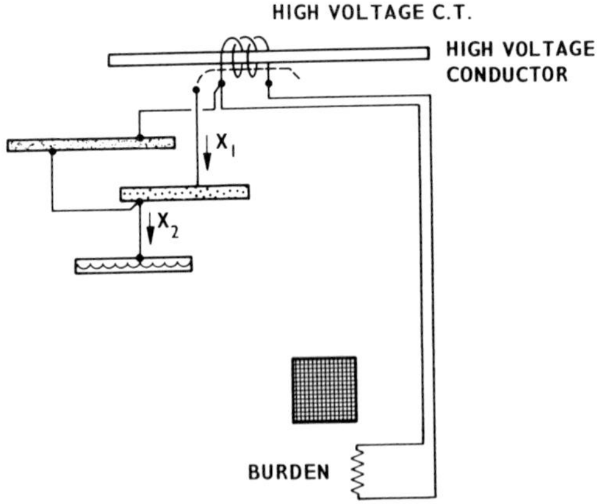

EARTHED NEAR NOISE SOURCE

FIG. 37

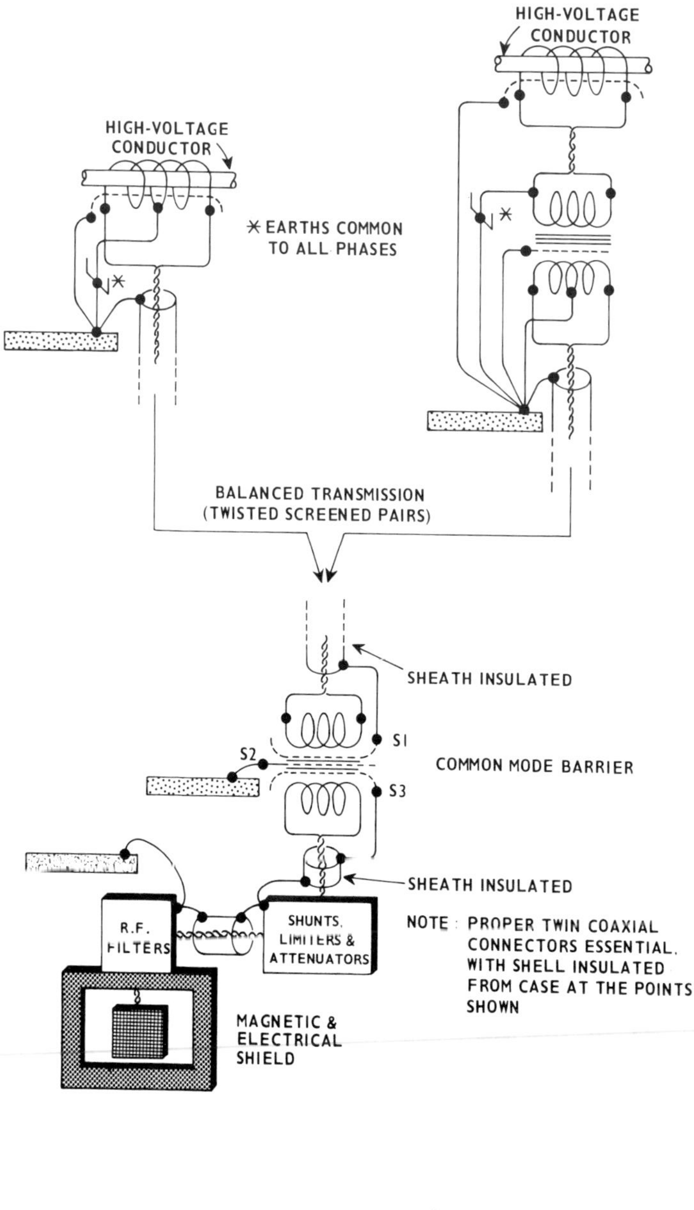

FIG. 38 CONNECTION OF CURRENT TRANSFORMER

terference into equipment in the very room where the vulnerable equipment is to be located.

Some values of interference levels due to capacitor voltage transformers (CVTs) are given in Chapter 5.

Fig. 37 illustrates the preferred arrangement whereby the current transformer secondary is *earthed at the H. V. C.T.* This confines interference currents to a smaller loop near the high-voltage source and must therefore reduce the noise levels in the apparatus room. Nevertheless the burden illustrated diagrammatically in Fig. 37 will still comprise a suitable Faraday-screened interposing current transformer. (The clean instrumentation earth shown in Fig. 37 is not necessarily common to the main clean instrumentation earth but is included in the diagram because a local bar will be a requirement for derivation of a common insulated point to which C.T. secondaries [not illustrated] from other phases can also be returned.) The common arrangement will be that the secondary of the interposing C.T. will drive a shunt, the potential across which will be fed through screened twin twisted wiring to limiter elements and thence, via any signal processing, to a radio interference filter at the entry into the screened area. Fig. 38 illustrates the preferred arrangement. It will be seen that the screening of the secondary wiring of the interposing C.T.s is earthed at one end only—again to confine return of interference currents to a loop remote from the vulnerable equipment. While appreciable attenuation of radiofrequency interference will occur in the considerable lengths of multicore cable normally found in practical installations, it is nevertheless desirable to minimise the area of loops and to provide means of shunting interference currents as near to their source as is possible.

As in the case of voltage transformers, it would be preferable to earth the *midpoint* of the C.T. secondary winding at the C.T. and to transmit balanced signals to the equipment, via an interposing common mode barrier transformer and filter assembly placed close to the vulnerable equipment. It should be noted that a physical midpoint (i.e., half the number of turns) may not necessarily be the electrical midpoint when considering a broadband radiofrequency case on account of unbalances in stray capacitances and residual inductances (see Fig. 38).

Whilst the introduction of balance/unbalance transformers may provide some benefit where existing cables have to be used, Fowler has pointed out (Appendix B) that superscreened cables provide a more effective solution than the employment of balanced transmission.

5 Interference Measurements in Substations

Extensive studies of induced voltages have been made by a number of workers, and many results were given in the IEE Conference on Electrical Interference in Instrumentation (IEE, 1970).

Birch, Burrows and Turner (1968) and Birch and Turner (1968) published information on the amplitude of voltage transients occurring on CVT, CT and carrier coupling circuits associated with 275 kV stations.

CIGRE Conference Paper 36-09, 1976 (Anders and Campling, 1976) provides useful data of measured common mode interference voltages and also of the effectiveness of their reduction by earthing both ends of the auxiliary cable sheath. A summarising table from that paper is reproduced below.

Most of the measurements did not provide conclusive evidence of the mechanism by which these common mode voltages were imposed on the auxiliary wiring, though there is little doubt that the worst conditions result from transient voltage rise of the earthing mat at capacitor voltage transformers (CVTs) during switching operations (particularly with isolators). A CIGRE Conference Paper 36-05, 1970 (Borgvall, Holmgren, Sunden, *et al.*) provides a comprehensive study of this problem, and a Figure from this paper is reproduced in Fig. 39 showing the magnitude of the high-frequency transient voltage rise at various locations in a substation. Feist

Table 2: Selection of Measured Results of Typical Interference Voltages in Auxiliary Cables During Switching Operation in H.V. Substations

				Common Mode Voltage					
Item	Initiating Device and System Voltage (kV)	Type of Auxiliary Cable and Length (Meters)	Type of Cable Screen	Without Screening (Volts-Peak)	With Cable Single-End Earthed (Volts-Peak)	With Cable Double-End Earthed or with Reduction Wire (Volts-Peak)	Reduction Factor	Frequency Range of Components (kHz or MHz)	Remarks
1	Isolator switch operation (150 kV)	Multicore 100 m	Steel foil over-all plus continuity wires	900	—	105	0.12	90 kHz 180 kHz	
2	Circuit-Breaker operation (380 kV)	Multipair	Braid over-all	2,900	—	146	0.05	300-400 kHz 1.0-1.7 MHz 3.0-3.8 MHz	
3	Circuit-Breaker operation (420 kV)	Multipair	Braid over-all	9,300	—	93	0.01	290-370 kHz 3-9 MHz	
4	Circuit-Breaker operation (380 kV)	Coaxial	Steel tube coax plus braid over-all	4,400	—	4	0.001	300-400 kHz 1.0-1.7 MHz 3.0-3.8 MHz	
5	Switching operations (40 kV)	25-100 m	Unscreened	2,000-6,400	—	200-300	0.1-0.05	1.5-2.5 MHz	Screening was afforded by a copper tube laid in the trench adjacent to the auxiliary cable. Circuit earthed at C.T.
6	Switching operation (130 kV)	50 m	Copper wire with "loose" coupling	10,000	600-1,000		<0.1	1.0 MHz	Circuit earthed at C.V.T.
7	Switching operations (400 kV)		Lead sheath	16,000	—	100	0.006	100 kHz and 1.0 MHz	
8	Switching operations (400 kV)	200 m	Copper wire with "loose" coupling	20,000	—	200	0.01	100 kHz	Circuit earthed at C.V.T.
9	Isolator switch operation (400 kV)	Multicore 17 m	Steel wire armour		20,000	700	0.035	1 MHz	Circuit earthed at C.V.T. screen earthed at opposite end for single-end earthing.
10	Power cable switching (10 kV)	Coaxial 10 m	Copper-braid	110		0.6	0.006	2 MHz	

(Reproduced by permission of the Authors and CIGRE)

(1977) considered the special problems of earthing in sulphur-hexafluoride switching stations.

The 100% reference value was taken at the CVT and was up to 8.2 kV during the tests, its value depending on a number of factors affecting the test conditions. (Other tests, e.g., those referred to in CIGRE Paper 36-09, 1976, indicated that transient voltages of 10-20 kV may occur in extreme cases). Fig. 39 shows that transient voltage rises in the kV range may be found even at distances of say 50-100 m or so from the CVTs. Both CIGRE papers also show that these common mode voltages imposed on the conductors can be reduced to a very small fraction of their original value by providing a low-resistance screen on the cable earthed at both ends. It should be emphasised here that the circuit itself should not be earthed at more than one point, and if the outer conductor of a coaxial cable is part of the signal circuit then this should not be earthed at both ends (even though it is often referred to as the screen). However, a further separate screen (often referred to as double screening) can be very effective in reducing common mode transient voltages.

The common mode power-frequency voltages imposed on the auxiliary cabling of substations are in general of lower amplitude than those containing high-frequency transients. A maximum value of about 500 V rms was estimated for a typical 400 kV substation at 50 kA earth fault current (Anders and Campling, 1976).

As circuits can never be perfectly balanced, particularly at higher frequencies, large common mode voltages produce some transverse voltage. Usually this cannot be expressed as a simple percentage of the common mode voltage as several factors influence its value. Though in most practical cases encountered at substations, earthing of the auxiliary cable sheath at both ends reduces not only the common mode but also the transverse voltage (particularly if the circuit is not well-balanced), there are also cases where the position is the reverse (see Section 2.5 of the CIGRE paper by Anders and Campling).

As transverse voltages are affected by more factors than common mode voltages and also because they are more influenced by design dctails, published literature provides relatively little numerical data and the establishment of the

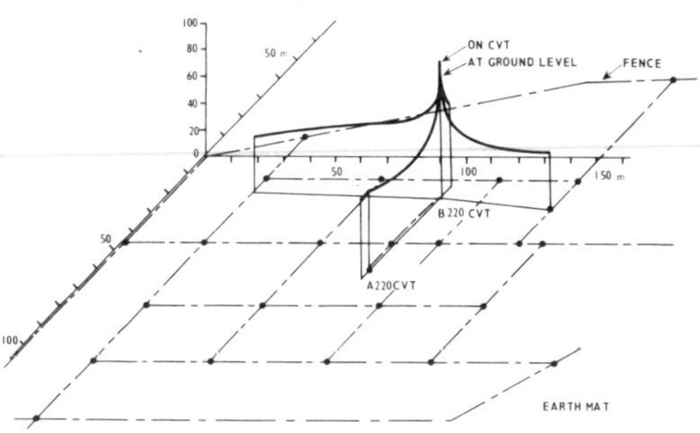

(BORGVALL, HOLMGREN, SUNDEN et al. C.I.G.R.E. PAPER 36-05, 1970 REPRODUCED BY PERMISSION)

FIG. 39 H.F. TRANSIENT VOLTAGE RISE AROUND CVT

mechanism involved (whether they are imposed by magnetic or electric fields or by voltage differences) is even more difficult than in the case of common mode voltages. In considering the problem, it has to be borne in mind that, within, say, a 400 kV substation, the transient voltage difference between various points of the earthing may reach values of several kV, and stray capacitances between various equipments of an electronic scheme and the nearest parts of the earthing may thus be a source of interference, a problem which does not exist to the same extent in power stations.

In the past, interference problems were virtually restricted to cases where the interfering signal was picked up by the auxiliary cabling. With the rapidly increasing use of sophisticated computers and electronic devices, the magnetic and/or electric field in a substation might adversely affect the correct functioning of these devices irrespective of the cables connected to them.

The 50 Hz magnetic field caused by a given fault current can be assessed by calculation if all the physical dimensions are known. It can also be measured by using a search coil.

A circular search coil of mean diameter d is considered, with N turns forming a winding of depth a and width b, both a and b being small when compared with d. A nonmagnetic former (insulating material, wood or aluminium split to avoid a shorted turn) is used. The magnetic field strength H is assumed constant across the coil.

A magnetic field strength (normal to the plane of the search coil) $H(Am^{-1})$ causes a flux density of:

$$B = \mu_0 H \text{ tesla}$$

where $\mu_0 = 4\pi.10^{-7}(Hm^{-1}) =$ permeability of free space.

The flux enclosed by the coil is $B\pi d^2/4$, and the flux linkages are $NB\pi d^2/4 = H \times \mu_0 \pi N d^2/4$. The instantaneous voltage induced by the changing field is then:

$$e(t) = \frac{dH(t)}{dt} X \mu_0 \pi N d^2/4 \text{ volt}$$

Specifically, if H varies as $H(t) = H_{max} \sin \omega t$, then the peak voltage measured is

$$e_{max} = \omega H_{max} \mu_0 \pi N d^2/4$$

At 50 Hz a field of 500 A_{rms}/m induces an e.m.f. of 0.2 mV_{rms} in a loop of 10 cm^2. At 5 MHz a field of 5 A_{rms}/m induces an e.m.f. of 0.4 V_{rms} in the same loop.

An alternative method of magnetic-field measurement, and, subject to bandwidth considerations, one that offers advantages where a measuring probe may need to be small enough to be used within an apparatus box to explore the internal field, is the Sony "magneto diode" type SMD (Entry 29, Appendix A). Similar devices are offered by other makers (Entries 10, 58, Appendix A). Such sensors may be operated with a broadband transient memory voltmeter to store the peak value of the magnetic field measured during, say, an isolator operation. Where the spectral content is required, this could be obtained by inserting one of several appropriate selective filters between the transducer and the memory voltmeter and making a series of measurements. An accuracy of ±10% would be more than adequate but may be difficult to achieve. Precautions are necessary to avoid any unwanted interference ingress to the instrumentation itself by direct RFI or supply conduction. Such an instrument would need to have a known response to transient as opposed to steady state fields, and where loops are used shock excitation of tuned circuits would have to be allowed for.

Current-monitoring devices which appear suitable for the measurement of currents in earth conductors, cable sheaths, etc., are offered and are claimed to cover the frequency range 1 Hz to 35 MHz and to give a typical pulse amplitude accuracy of +1% −0% with a 10 ns risetime (Entry 50, Appendix A).

During a power-system fault, the magnetic field will be decisive principally at 50 Hz; but during the onset of the fault and when it is being cleared there may be intense high-frequency components extending to tens of MHz. As already pointed out, during isolator operation intense broad-spectrum interference extending to tens of MHz is produced; and although very few specific measurements have come to light, it is known that high levels exist. Both the magnetic and electric fields may be of interest.

Laborelec (see Section 4.1) have considered a worst-case 50 Hz field of 500 A_{rms}/m corresponding to a 50 kA earth fault on their power system (see also Chapter 6). Information on the spectrum and level of magnetic fields during faults, isolator and switching operations is not readily obtainable. In the absence of measurements, assumptions could be made to derive the peak currents flowing and to estimate their spectral content for any particular busbar or substation configuration. Whilst values of 25.5 and 37.5 A_c/m are reported in U.K. (see Section 4.1), these can be assumed to be at low frequencies. There is thus a requirement for a series of measurements to determine the order of the level and spectral content of fields produced during isolator and switchgear operation, particularly in the range above 50 Hz extending to at least, say, 20 MHz and preferably to 100 MHz. Such measurements could explore the spectrum and level, not only of the direct fields at the equipment but also of the currents flowing in isolator support earth strips and also in CVT and CT earth strips during isolator operation (see Chapters 10 and 11).

So far in the discussion, reference has been made to magnetic and electric fields separately, because at the predominant frequencies and also the physical dimensions involved, the interference produced by the high-voltage primary system is not of the genuine radiated type and that kind is not likely to be the main source of possible trouble. The 50 Hz and, in general, also the transient magnetic and electric fields are not directly related to one another. Each necessitates specific measures, and the requirements are not the same for the two cases; e.g., very effective electric-field screening at 50 Hz may be useless against magnetic fields of the same frequency. In the higher MHz region where the high-voltage system may represent an efficient radiating antenna, there is a firm relationship between the magnetic and the electric field.

For an electromagnetic wave propagating in free space and in the Z direction with magnetic and electric field strengths of Hx and Ey, respectively:

$$\mathrm{Ey} = \sqrt{\frac{\mu_0}{\epsilon_0}}\,\mathrm{Hy}$$

where ϵ_0 is the permittivity of free space ($\sim 8.854 \times 10^{-12}$) and

$\sqrt{\frac{\mu_0}{\epsilon_0}}$ is the intrinsic impedance of free space and has a value in MKS units of $120\pi \cong 377$ ohms.

Thus the measurement of the magnetic field in this case provides the value of the electric field as well and can be measured by a search coil. The electromagnetic field is measured conventionally by the use of three search coils mutually at right-angles, placed at the location where it is desired to measure the field strength. From the voltages induced in the search coils, the three components can be derived. If these are denoted by H_x, H_y and H_z, then the resultant magnetic field strength is simply:

$$H = \sqrt{(H_x^2 + H_y^2 + H_z^2)}$$

As an example, take:

$$d = 0.1 \text{ m}$$
$$N = 200$$
$$H_{max} = 250 \text{ A m}^{-1}$$
$$W = 314 \text{ rad s}^{-1}$$
$$e_{max} = 314 \times 250 \times 4\pi 10^{-7} \times \pi \times 200 \times 0.01/4$$
$$= 0.155 \text{ volt}$$

It will be clear that a sensitive detector will be required. The detector must present a high input impedance to avoid loading the search coil, and must of course be totally unaffected itself by the magnetic field. This may involve orienting an oscilloscope, for example, so that the field is aligned with the Z-axis of the cathode-ray tube.

If desired, the search coil may be tuned by a capacitor, connected either in parallel or in series. The output voltage e_{max} will then be multiplied by the Q of the resultant tuned circuit. However, this is only applicable at the given resonant frequency and for a steady-state signal.

Alternatively, if a wideband response is desired, the coil

can be loaded with a suitable resistance such that the frequency response is levelled.

It should be noted that:

$$\begin{aligned}\text{Magnetic field strength } 1\ \text{A m}^{-1} &= 4\pi 10^{-3}\ \text{oersted}\\ &= 1.2566\times 10^{-2}\ \text{oersted}\\ \text{Magnetic flux density } 1\ \text{tesla} &= 10^{4}\ \text{gauss}\end{aligned}$$

In free space (or air), a field strength of 1 A m^{-1} corresponds to a flux density of $4\pi 10^{-7}$ tesla $= 4\pi 10^{-3}$ gauss.

6 Susceptibility of Computers to Interference

Extensive work by Laborelec (see Section 4.1) (Laborelec [a]-[j] inclusive, Bibliography; and private communications) has been directed to gaining an appreciation of the order of magnitude of interference to which certain common types of computer can be subjected before operation is corrupted.

Because the electromagnetic screening of many commercial computers is often rudimentary, its effectiveness in certain machines has been investigated, over a frequency range extending from 50 Hz to 4 MHz, in three mutually perpendicular planes, by means of a search coil placed within the cabinet, the whole of the small computer and its peripheral being placed in a suitable external solenoid. The effectiveness of the screening varied considerably with frequency and the plane of the applied field and may on average reduce the internal field to only about half of its external value. It is known that investigations at much higher frequencies would also be useful, but limitations of the test equipment prevented this.

Insofar as 50 Hz magnetic fields are concerned, levels of up to 500 A_{rms}/m were applied for 30 seconds; 500 A_{rms}/m corresponds to the maximum value expected in a particular Belgian substation in the case of a single-phase 50 kA fault on the line at the first transmission tower with the computer building situated directly below the line. The computer functioned normally.

Tests were made in which a series of damped high-frequency oscillations (having a damping time constant of about three times the period of a full wave at the frequency concerned) was injected in the external coil using frequencies between 400 kHz and 4 MHz. Ideally, the frequencies used should have extended to much higher values, but limitations in the available source fixed the upper frequency at 4 MHz. During these tests, the computer casing was first removed and then left in position, and the three planes of the loop were again used together with a search coil oriented for maximum coupling and measuring the internal field. The test durations were one minute at a repetition rate of twice per second (limited by the generator). The computers were run with various test programs and were operated either without peripherals or with a teletypewriter. At the start, the computer was working alone without peripherals so that it could be certain that no interference could enter by way of the connections. Later the typewriter was added, using a short shielded cable. The connection of the typewriter had no effect on the sensitivity level measured. No input was connected, and the supply was filtered and measurements were made to prove that interference was not entering by the supply leads. The purpose was to analyse the computer sensitivity to *radiated electromagnetic interference.* The results corroborated the fact that interference was entering by radiation: the sensitivity of the computer was appreciably increased when the metal screening was removed. The computer cases were of the normal manufacturers' types, and for the particular tests ingress of interference by means of peripheral connections, power supplies and earth circuits can be discounted.

Table 3, reproduced by kind permission of Laborelec, gives some typical results for a particular computer. H_o is the field in the environment in which the computer is placed and is the field produced within the solenoid by the test interference source in the absence of the computer. The introduction of the computer modifies the field distribution. H_i is the field measured within the cabinet at a point at least 10 cm away from metallic components. The central processing unit is itself placed in a further inner metal container where the field is appreciably lower than H_i. (At 1 MHz it is about 0.25 H_i.) Columns 2 and 5 in the Table give the maximum value of the ambient field produced in A peak/m.

The Table indicates the maximum possible values supplied by the test equipment at which computer operation remained normal or the lowest values at which the computer functions were disturbed.

Table 3

(1) Frequency MHz	Metal Casing Removed			Metal Casing in Place		
	(2) Ho(Ac/m)	(3) Hi(Ac/m)	(4) Finding	(5) Ho(Ac/m)	(6) Hi(Ac/m)	(7) Finding
			HORIZONTAL LOOP			
0.4	70	48	Perforator motor started	85	28	Functions correctly
0.6	50	32	Perforator motor started (A)			
0.85	40	26	Computer stopped (C)	63	18	Functions correctly
1.2	40	24	Computer stopped	60	14	Functions correctly
2.5				65	17	Computer stopped (C)
4	30	12	Perforator motor started	60	15	Computer stopped (C)
4	35	15	Loss of supply signalled (B)			
			VERTICAL LOOP PARALLEL TO FRONT FACE			
0.4	60	40	Functions correctly			
0.85	35	20	Computer stopped (C)			
1.2	21	14	Computer stopped (C)	25	9	Computer stopped (C)
4.0				15	6	Computer stopped (C)
			VERTICAL LOOP PERPENDICULAR TO FRONT FACE			
0.85	25	16	Computer stopped (C)			
1.2	20	14	Computer stopped (C)	15	6	Computer stopped (C)

(A) Tape punch motor started at wrong time.

(B) Power-loss failure signal given due to faulty operation of detection circuit.

(C) Shutdown of computer following signalling of voltage failure and faulty operation of the routine for saving the content of the registers in certain store addresses due to interference, falsification of addresses or errors in parity.

(D) Ac/m = Amperes crest per meter.

It is strongly emphasised that the values for the sensitivity threshold are orders of magnitude only. *There can be no question of taking them as absolute values.* Two tests under identical conditions produce results differing by as much as 30%, one reason being the correlation between the interference waveform and the execution of certain functions within the computer program.

The 50 Hz tests confirmed that the particular computer has a low sensitivity to low-frequency magnetic fields. However, in the range 400 kHz to 4 MHz the sensitivity increased with frequency. At 4 MHz the threshold was about 15 Ac/m, which is about the same as the field reported near 150 kV isolators. Further tests confirmed that the sensitivity further increases with increasing frequency; a field of only 8 Ac/m containing a high proportion of HF ($\cong$ 15 MHz) will shut down the computer. An arc from a 60 kV impulse generator at a distance of 10 m caused computer shutdown. In an unshielded switchgear bay in a 150 kV station, near isolating switches, the fields can be expected to cause a shutdown, the program probably being corrupted and thus requiring reloading.

Although shutdown is likely, the possibility of damage to the components under the effect of electromagnetic fields (as opposed to overpotentials applied via peripheral wiring) can be discounted. The theory that orders may be issued at the wrong time under the effect of an interfering electromagnetic field must be excluded because of the serial nature of the working process of the computer. However, where short links are used between two computers of a particular type, errors were reported with fields H_o of the order of 15 Ac/m at 1.2 MHz and 10 Ac/m at 4 MHz.

A different type of computer exposed to the same type of test as in Table 2 gave results shown in Table 4, from which it can be seen that the screening was far more effective than in the case of the computer of Table 2.

Table 4

(1) Frequency, MHz	(2) H_o (Ac/m)	(3) H_i (Ac/m)	(4) H_o/H_i	(5) Finding
0.3	400	20	20	Computer stopped
0.42	350	16	22	Computer stopped
0.55	300	12	25	Computer stopped
0.65	250	9	28	Supply failure falsely signalled
0.65	280	10	28	Computer stopped
0.9	250	7	36	Functions correctly
1.3	300	6	50	Functions correctly
2.0	240	4	60	Functions correctly
4.0	230	3.5	66	Functions correctly

The field H_o characterises the surroundings of the computer and is the field produced in the solenoid in the absence of the computer. The field H_i is measured inside the computer. The ratio H_o/H_i gives a value for the effectiveness of the screening of the computer. Column (2) gives the value of the ambient field at which the conditions of Column (5) apply or the maximum value which could be achieved with the test equipment in those cases where correct functioning still occurs.

It will be seen that as the frequency is increased the sensitivity threshold is decreased to a minimum of 250 Ac/m at about 650 kHz and then increases again (see also Chapter 7). (This can be contrasted with a value of about 50 Ac/m for the previous type of computer—see Table 3.) The sensitivity threshold increases at higher frequencies; and, because of limitations in the test source, it could not be determined above 900 kHz. The variation of the threshold with frequency is probably the result of two actions, namely the increase with frequency in the sensitivity of the computer to the internal field H_i and the increasing effectiveness of the screening (see column [4]) with frequency. Screening of the second computer is effective because of the compactness of the particular unit. The probability of encountering an ambient field of the order of 250 Ac/m—the minimum threshold of sensitivity of the computer in question—is extremely low.

It can thus be deduced that between the two computers examined there are wide differences of sensitivity to interfering fields and that careful choice of computer—or the provision of extra screening in appropriate cases—will be essential.

Computer shutdown and destruction of the program must be expected as a result of high-frequency fields if the computers are insufficiently screened or are placed in the switchgear buildings, and particularly if placed near isolators.

It is also apparent that proof tests should be required, whereby a computer manufacturer should demonstrate and state the maximum environmental fields in which the particular configuration can be successfully operated, so that these figures may be contrasted with projected field levels within substations, thus confirming the adequacy of screening measures (Chapter 10).

Further work has emphasised that computers are also most susceptible to high-frequency interference entering by the supply. Tests in which a 1/50 μs impulse of 1 kV was applied between the supply phase and earth were made. The computer signalled a supply failure, and in one test a supply fuse was destroyed. Damped sine waves with a damping time constant of 3 periods and at a frequency of 300 kHz repeated twice per second from a source impedance of 200 ohms were also used. A 1,200 volt peak interference signal caused computer shutdown when connected between phase and earth. The strictest requirements will have to be observed in filtering and isolating the supply (Section 4.6.2).

Similarly, connections to the peripheral input and output systems are extremely vulnerable to the ingress of interference. However, means of limiting the amplitude of undesired interference voltages and means of filtering high frequencies are relatively easily applied (Section 4.5).

Laborelec consider (private communication, 22 December

1976) that a computer (for use in a high-voltage substation) must be insensitive to:

- a magnetic field at 50 Hz equal to 500 A effective/m.
- sustained electromagnetic fields equal to 2 V pk/m
0.2A pk/m
(in the range up to 200 MHz)
- transient electromagnetic fields equal to 50 V pk/m
5 A pk/m
(damped oscillatory wave and fast transient)

The Laborelec requirements concerning input, output and supply circuits depend on their individual functions which are distinguished in four categories:

1. Connections to voltage or current transformers.
2. Connections outside the buildings other than 1.
3. Within the buildings, unprotected circuits.
4. Within the buildings, protected circuits including screened circuits and unperturbed short connections situated far from interference sources.

For each category, Laborelec recommend a transient test and a test with a damped oscillatory wave of frequency 100, 200, 400, 800 and 1,600 kHz and a time constant of 3-6 periods with a source impedance of 50 ohms. The amplitudes applied in the four categories are as follows:

1. 2,500 volts peak between phases
between phases and neutral
between neutral and earth.
2. 1,000 volts peak between wires and earth
500 volts peak between wires.
3. 500 volts peak between wires and earth
between wires.
4. depends on circuit type.

7 Susceptibility of Solid-State Circuits to Noise and Damage

The failure mechanism results from localised heating and the necessarily limited thermal-transfer capability between the device itself (which is often of very small physical size) and the heat sink (Turner, 1968).

Clearly, devices designed for operation at high voltages will have greater interference immunity. High-noise-immunity logic systems with optoisolation have been shown to operate satisfactorily in substation switchgear testing conditions.

The differing design and operating method of the various families of device result in widely differing withstand capabilities. Because false operation of logic devices by ingress of interference occurs at a lower level than damage and because avoidance of false operation is an important criterion in the substation application, this chapter hereon refers only to interference immunity in respect of avoidance of false operation rather than to destruction of the device.

Clearly, many of the systems used in substations will employ proprietary equipment where the choice of the type of device has already been made. In other cases, however, for example where a new equipment is being designed to meet a specific requirement, there will be an opportunity to influence the choice of the devices so as to take advantage of noise immunity or other properties.

It is clear, for example, that choice of integrated-circuit families, insofar as interference immunity is concerned, favours the ranges using higher operating voltage levels (15 or 18 volts D.C.). However, the failure rate of this range (through internal breakdown, etc.) is slightly higher than that of the lower-voltage types. The subject is a large and specialised one, and it is not appropriate to discuss it in detail here; but as an example the following figures published by Motorola relate to their CMOS family (Reliability Report, CMOS, 1975). Based on long-term life test data, the current 1975 failure rates are as shown in Table 5.

Table 5

Device Series	Operating Voltage	Failure rate % per 1,000 hours at 60% confidence level	
		85° C	125° C
AL	18	0.0150	0.432
	10	0.00092	0.028
CL	15	0.076	*
	10	0.0132	*
CP	15	0.120	*
	10	0.009	*

*Exceeds data-sheet operating-temperature limitation.

When selecting a logic family, one has to consider that to produce a false signal on a line connecting two circuits the interference source must be able to generate a large enough current to force the input above the switching threshold for long enough for the false signal to propagate through the following circuit. One therefore has to consider the energy required to force a change of state. As the duration of the interference pulse is decreased it becomes more nearly equal to the gate response time (which depends on its type) and the voltage needed to trigger the gate increases. A plot of energy interference immunity against interference pulse width has the shape illustrated in Fig.

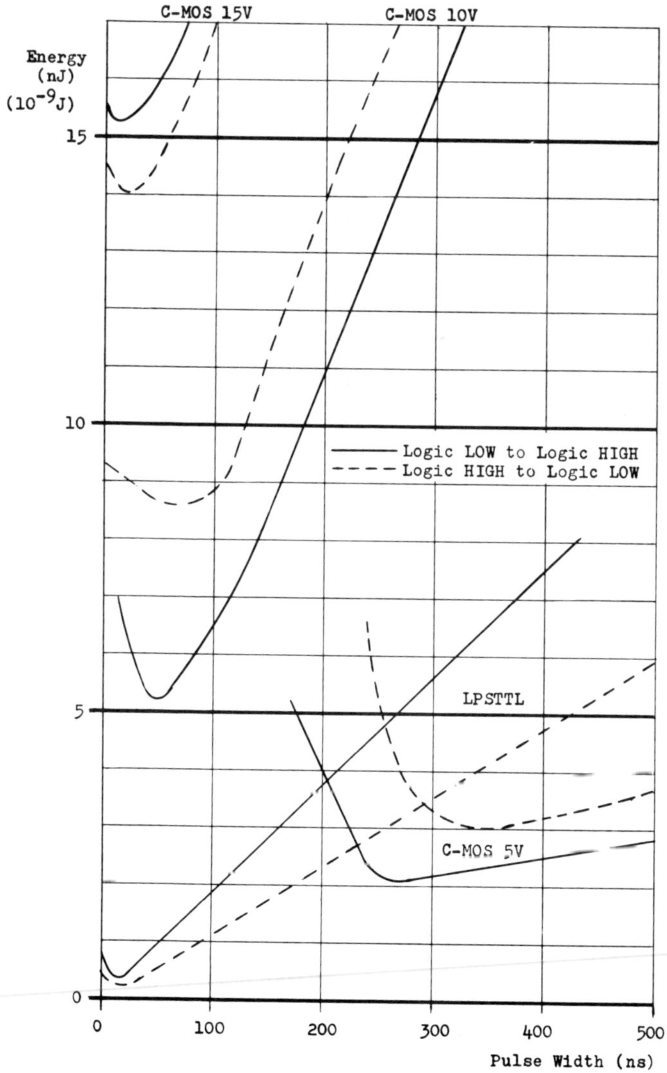

FIG. 40 ENERGY INTERFERENCE IMMUNITY OF TYPICAL INTEGRATED CIRCUITS

(taken from: BISHOP, A., "Comparison of low power Schottky TTL and C-MOS", New Electronics, 15 October 1974, and reproduced by permission of the Author.)

40, and the energy noise immunity is a product of the interference voltage, interference current and pulse width. A gradually decreasing power interference immunity occurs as pulse width is reduced, until a minimum is reached. At this point the curve rises steeply because the gate responds less to narrow interference pulses. The energy interference immunity at the minimum point is an appropriate value for comparison of different logic families. It should be noted that the *minimum immunity typically occurs with pulse durations in the range 10-100 ns (see also Table 6), corresponding to frequencies of 10-100 MHz, thereby emphasising the requirement for adequate attenuation and filtering to prevent ingress of such frequencies.* The interference immunity for LPSTTL and CMOS (see Glossary, Section 3.1) has been stated (Bishop, 1974) to be as shown in Table 6.

Table 6

	LPSTTL	CMOS
Power supply	4.75-5.25 V	3-15 V
Interference immunity worst case:		
5 V supply	0.3 V "1" level 0.7 V "0" level	1.5 V both levels
15 V supply	nonoperable	4.5 V both levels
Propagation delay $C_L = 15$ pF		
5 V supply	10 ns	35 ns
15 V supply	nonoperable	18 ns

The CMOS handbook published by Motorola gives detailed information on the subject; and this is summarised in Table 7 below, abstracted from that document.

Table 7

Logic Family	Power Supply (Volts)	Typical Gate Quiescent Power Dissipation (mW)	Typical Propagation Delay (ns)	Typical Signal-Line D-C Noise Immunity (Volts)		Typical Signal-Line Impedance (ohms)		Logic Voltage Swing (Volts)	Typical Energy Noise Immunity on Signal-Line (Joules 10^{-9})	
				Low	High	Low	High		Low	High
DTL	5	8	30	1	3	50	1.7K	4.5	3	1.5
TTL	5	15	10	1	3	30	140	3.5	4	2.5
HTL	15	30	85	4.5	6	140	1.6K	13	48	7
ECL	−5.2	25	2	0.25	0.17	7	7	0.8	N.A.	N.A.
CMOS $V_{DD}=5$ V	5	5.10^{-3}	45	2.25	2.75	600*	1.2K**	5	3	1.5
CMOS $V_{DD}=10$ V	10	10.10^{-3}	16	4.5	5.5	300*	600**	10	10	5
CMOS $V_{DD}=15$ V	15	15.10^{-3}	12	6.75	8.25	250*	450**	15	22	13
CMOS $V_{DD}=18$ V	18	18.10^{-3}	11	8.1	9.9	220*	430**	18	35	18

*Defined at 30% of V_{DD} } D.C. resistance
**Defined at 70% of V_{DD} }

It should be noted that there are many other considerations such as price, second sourcing, availability, power dissipation, speed, crosstalk and power-supply requirements which must be taken into account when selecting an IC family for a given application. The reader concerned with these questions and with interference immunity is recommended to study the literature (Allan, Bishop, Blood, Maul and papers referenced therein; also see Bibliography, Chapter 15).

Whilst ideally the choice of devices can lead to better immunity to interfering voltages, it has to be noted that such a recommendation alone is misleading and dangerous. This is because:

(a) Many disturbances are internally generated, and higher-immunity devices have higher levels of generation; thus crosstalk immunity is not affected.

(b) The higher-immunity devices are more susceptible to damage; they absorb a higher proportion of the total disturbing power in a well designed system.

(c) It is undesirable to recommend particular families of devices because of the rapidly evolving state of the art and the need to give the circuit designer freedom of choice.

Clearly, viewed from the point of view of interference immunity alone, 18 volt CMOS is a likely choice; but much depends on the circuit design itself. (For definitions of the abbreviations, see Glossary, Section 3.1.)

A more detailed discussion of the topic is given by J.F. Dixon in Section 11.14 of a report on microprocessors and minicomputers (Dixon).

Note: 1 Joule $= 10^7$ erg $= 2.77672 \times 10^{-7}$ kWh
1 erg $= 10^{-7}$ Joule
1 watt $\cong$ 1 Joule/second

8 Typical Values of Attenuation Required

Because of the comparative scarcity of measurements in substations of the amplitude and even more so of the spectrum of actual interference on a statistical basis, the best judgment that can at present be effected must be based on experience in the operation of the many existing equipments. Large numbers of electronic systems, many of which may be considered vulnerable, are already in service in high-voltage stations. In most cases, the systems operate satisfactorily; but in other cases some malfunctions have been manifest. By means of laboratory measurement of the susceptibility of samples, it would be relatively simple to set an appropriate level above which systems could be expected to operate correctly with high confidence. Correspondingly, any new device or system proposed for installation could be evaluated; and, if its susceptibility were measured as being below the set level, it could be expected to malfunction in a service environment. Any necessary remedial measures could be applied and the improvement in the susceptibility level measured.

If a standard broadband interference susceptibility test, such as that of UKAEA (Section 10.11), were applied as a type test to each existing vulnerable plant item, then it would be possible to build a dossier in which listed plant items were arranged in order of their interference-withstand abilities. The experience obtained by the exposure of many examples of each item to the real environment as a result of their service within substations provides a large sample from which meaningful results are already obtained (when a given equipment is never affected, is occasionally affected or is constantly in trouble).

This process is likely to be more meaningful than the measurement of specific conditions within a few substations and to yield results more quickly.

Because of the present lack of an agreed method of effecting a broadband susceptibility measurement and the nonavailability of the proposed dossier, no final specific values of required attenuation can yet be given.

The need to derive expressions linking far-field radiation levels to currents flowing in a conductor as suggested by Fowler (Appendix B, Section B.5) is also a clear requirement.

8.1 Voltages Entering by Conduction

The sensitivity of a particular equipment can be defined (Section 10.11) from an examination of its input/output and power-supply circuits and a consideration of the factors given in Chapter 7. Alternatively, the equipment manufacturers could state maximum tolerable common and series mode interference levels at the equipment terminals, plotted against frequency.

Knowledge of the likely maximum voltages (Birch, Burrows and Turner, 1968; Birch and Turner, 1968; Turner, 1968; Anders and Campling, 1976) arising in substations gives some orders of magnitude of the attenuation required.

Since many solid-state systems are using signal levels of the order of a few volts and overvoltages are of the order of several kilovolts, attenuations exceeding 60 dB will be needed. This is readily achieved by barrier devices and limiters. Because not much is known quantitatively of the spectrum of the interference, such knowledge could be strengthened by measurements at typical sites of both voltage and fields (see Chapter 5). Present indications are that a safe precaution would be to follow the barrier attenuator limiter assembly by an R.F. filter having a typical attenuation of 40 dB over the range 100 kHz to 100 MHz. Where high speed of response is not a prime requirement, then this filter may constitute an R.C. low-pass network. Attention is drawn in this case to the need for careful choice of the capacitors to obtain a low R.F. impedance at the higher frequencies. Conventional capacitors are often highly inductive and will need to be replaced or shunted with R.F. types.

If common mode interference occurs, the main requirement is one of insulation; so that there may not always be any need for attenuation of this mode. However, in the case of a series (transverse) mode interference, then the required signal-to-noise ratio is decisive in the choice of the required attenuation.

Where choice exists, high-noise-immunity circuits should be specified having regard to the comments in Chapter 7.

Conducted interference should also be guarded away from the vulnerable equipment by the use of balanced transmission, screened cables, Faraday screening, barriers and single point earthing. Also, attention should be given to the points made in this book.

In the UKAEA, it is considered that an instrument cable screen should be capable of supporting 100 mA of interference current in the frequency range 10 kHz to 100 MHz, and 300 mA to ensure immunity from thunderstorms. However, the high-voltage-substation case is clearly more difficult than that of a nuclear power station; so higher values may well be required.

Bull (1975) has suggested minimum values of series mode attenuation for power supplies to digital equipment generally (not necessarily in a hostile substation environment). An overall attenuation including mains and power unit output filtering is proposed as follows:

Frequency, MHz	0.3	1	3	10	30	100
Attenuation	3.10^4	10^4	3.10^3	10^3	300	100
dB	90	80	70	60	50	40

8.2 Effects of Radiation

Current evidence is that mini- or micro-computers and similar apparatus in a substation should be effectively screened against electromagnetic radiation. Also, in the case of typical vulnerable equipments placed near to busbars, attenuation for electric and magnetic fields should be at least as follows:

100 kHz -500 kHz	20 dB minimum
500 kHz - 2 MHz	30 dB minimum
2 MHz-100 MHz	40 dB minimum

These figures are tentative and may need to be restated when more is known quantitatively of actual fields and the susceptibility of the vulnerable equipments.

Lower attenuations may be sufficient in the case of equipment mounted in buildings further from the interference source; but nevertheless the screening will still be essential, to protect against R.F. fields caused by switching of inductive circuits (trip coils, contactors, motors, etc.) and by local radio transmitters (hand portable, base station and mobile radios).

It is not difficult to achieve an attenuation of at least 60 dB over the frequency range extending to 100 MHz by simple enclosure, providing that attention is paid to sealing and filtering of ingress leads in the ways set out in this book.

Where choice exists, the equipment enclosures should be arranged to provide good attenuation. In other cases, external enclosures (such as a modified equipment cubicle) will of necessity be a requirement as an external screen around the vulnerable equipment.

Attenuation of screening enclosures can be measured by means of an internal search coil using broadcasting stations as a source, or more properly by enclosing a tunable source within the enclosure and measuring the external field at a distance whilst rotating the enclosure through 360° on a turntable (ERA).

9 Check List of Desirable Precautions

9.1 Radiated Interference

(a) Adequate screening enclosure(s), preferably double, around vulnerable equipment.

(b) Minimum volume of screened enclosure.

(c) Sheet metal or foil.

(d) All edges jointed.

(e) No apertures.

(f) Modifications to conventional cubicles.

(g) Cubicle floors filled in.

(h) Earth strips outside.

(i) Buswires filtered.

(j) Enclosure earthed at one point, minimum capacitance to local earths.

(k) Only one bond between inner and outer enclosures.

(l) Cables enter at one point where earth strip joins.

(m) All cables screened with glands/shell connectors.

(n) Cables matched where possible.

(o) Conducting gaskets or contact finger strip round all doors.

(p) Ventilation through waveguides below cutoff frequency or gauze.

(q) Attenuation of at least 100 dB in frequency range 1 MHz to 100 MHz to be aimed at.

(r) Choice of commercial electronic equipment having thorough screening wherever possible.

(s) Reduction of interference at source wherever possible. (Incorporation of quench circuits on contactors, solenoid valves, fluorescent lights, etc.)

(t) Use of high-interference-immunity semiconductor devices.

9.2 Conducted Interference

(a) All cables screened, preferably solid metal. Avoid parallelism of power and signal circuit cables.

(b) Screens bonded by glands or shell connectors at entry and contiguous with enclosure.

(c) Where justified, all transmission should preferably be made using balanced twin twisted pair or superscreened cables.

(d) Signal and power cables laid to minimise coupling with station earth bars and high-voltage conductors insofar as is possible at 90°.

(e) Signal cables as short as possible.

(f) Electrostatic guard screens within electronics where possible.

(g) Clean instrumentation earth, common to earth mesh at appropriate point.

(h) Single point earthing within electronics, short earth leads.

(i) Barrier devices having high common and series mode rejection, to include Faraday screens, including vulnerable equipment power supplies.

(j) Choice of best form of signal; use high-level signals if necessary attenuated close to computer input. Signals should be dissimilar from the most commonly expected form of interference.

(k) Signal conditioning by filtering where possible. Use of high-level signals where possible.

(l) Barrier device screens separately and individually earthed.

(m) Avoidance of long wraps in wire wrapping techniques used on earth wires.

(n) Use of electrostatic screening earth planes in printed-circuit card circuit design and construction.

(o) Use of high-interference-immunity semiconductor devices.

10 Test Specifications and Methods

There are a number of types of test which are useful in assessing the susceptibility of individual items of equipment, and the adequacy of the screening, filtering and the earthing arrangements. Before briefly describing a few typical examples, we can subdivide them as follows:

1. Voltage-withstand tests	**Relevant Section**
(a) I.E.C. (TC 41) test	(10.1)
(b) BEAMA test for electronic relays	(10.2)
(c) IEEE surge withstand capability test	(10.3)
(d) Laborelec test	(10.4)
(e) Ontario Hydro transient immunity test	(10.5)
2. Mains interference tests	
(a) BSS.4509 Part 2, 1970; BSS.4525, 1970	(10.6)
(b) Beckman line noise generator	(10.7)
(c) Shaffner equipment	(10.8)
(d) Power line disturbance monitoring	(10.9)

3. Electromagnetic field withstand test for vulnerable equipment

(a) Following the method of Laborelec (Chapter 6), but enhanced to cover wider frequency spectrum and higher field capability (10.10)

4. Screening, barrier isolation and filtering tests

(a) UKAEA interference immunity tests (10.11)

The tests listed in the first and second group (and other similar tests not listed) are generally useful in assessing the quality of the hardware in standing up to overvoltages and in showing that the barrier isolation, limiting and filtering devices are functioning properly. However, because they do not necessarily expose the effects of resonances, they have only a limited value in revealing any shortcomings in the complete installation and do not show up deficiencies in interchannel coupling, in screening or in the frequency response of an assembly. Nevertheless, they are essential to prove the properties of the individual components. The tests listed in 2 above are important in exposing shortcomings of any equipment in respect of mainsborne noise and short mains interruptions. In particular, the use of the BSS or VDE tests (2[a]) is recommended. They are not described in detail, since they are well known and their choice presents no special problems.

The tests listed in 3 above are invaluable in exploring the effect of intense radio and power frequency electromagnetic fields on a computer or other vulnerable device. At present such tests are in an early stage of evolution, but it is already apparent that the method is a powerful aid in exposing shortcomings in the screening of a computer and can give an order of magnitude of the peak fields (and their frequency) which if exceeded can cause the machine to malfunction. Such tests could well be further developed and could advantageously become a requirement such that a computer (or device) manufacturer would have to state the maximum tolerable level in the environment just as he now quotes, for example, the temperature conditions. Clearly, more quantitative appraisals of the withstand levels of the types of equipment used will be needed, so that the figures can be used to quantify the amount of attenuation by screening which would be necessary.

One of the most useful test methods (4, above) is that developed by the UKAEA Reactor Group, Control and Instrumentation Division. This enables the interference immunity of individual items to be explored and the whole system to be progressively added so that the adequacy of screening and immunisation can be quantitatively established. It is clear that the method is ideal for use in substations, the more particularly since it is thoroughly documented and can be applied with confidence using commercial test apparatus.

Brief notes on some of the tests listed above are given below for reference purposes:

10.1 I.E.C. Test

The International Electrotechnical Committee No. 41 (IEC TC41) Working Group 4 define a high-frequency disturbance test in which a damped oscillatory wave of 1 MHz at a pulse-repetition frequency of 400 per second and having a 2.5 kV peak value for the first halfcycle (1 kV in the transverse mode) is applied for two seconds. The source impedance is 200 ohms, and for certain cases alternative levels of 1 kV and 0.5 kV are specified (see also IEC Publication 60 and Section 10.8[e]).

The test, although devised mainly for relays, can be regarded as being useful for the testing of the input barriers, limiters, filters, etc., in systems such as those with which this book is concerned.

10.2 The BEAMA Test

BEAMA (British Electrical and Allied Manufacturers' Association) have defined a test for electronic relays (BEAMA,

1966) in which no damage shall occur when a 1/50 μs wave of 0.5 Joule and 5 kV peak amplitude is applied, the shape being as defined in BSS 923:1940. The test voltage is applied to all combinations of terminals and earth of the unit and is intended to prove that the equipment will not be *damaged* by transient overvoltages. This test is used by the C.E.G.B., T.D.C.D. Protection Control and Measurements Group. It can be considered as comparable in form to that of Section 10.1, although a higher voltage is specified (see also IEC Publication 60). The test, originally intended for relays, is clearly applicable to testing the input barriers, limiters, filters, etc., in systems such as those with which this book is concerned.

A test set complying with BEAMA publication 219 is made by G.E.C. (A.E.I.) Ltd.

10.3 IEEE/ANSI Test

The IEEE formed a Working Group on static relay surge protection which produced a test procedure known as the IEEE Surge Withstand Capability Test. This is not greatly different from that of the IEC and uses an oscillatory wave of 1 to 1.5 MHz having a 2.5 kV to 3 kV peak value for the first halfcycle. The envelope is specified to decay to 50% of the crest value in not less than 6 μs. The repetition rate is 50 per second for not less than 2 seconds from a source impedance of 150 ohms (see IEEE Guide for Surge Withstand Capability Tests ANSI C 37,91a-1974, IEEE Std. 472-1974, and Electronic Apparatus in Control Equipment for Power Installations, Swedish Standard SEN 3610 503).

10.4 Laborelec Interference Tests

These consist of oscillatory damped wave tests at frequencies extending from 100 kHz to 2 MHz, and a fast transient test. Details are given by Jacquet (1977).

10.5 Ontario Hydro Transient Immunity Test

A type test using a damped oscillatory waveform with one

or more oscillation frequencies between 0.1 and 2 MHz is used. The peak instantaneous open-circuit voltage is up to 5 kV, falling to half amplitude in not less than 5 cycles. The transient is to be applied at a repetition rate of between 60 and 400 per second for 4 seconds from a source impedance of between 20 and 200 ohms (Harvey, 1977).

Table 1 of Harvey's paper also gives most valuable information on the worst-case transient levels for shielded and unshielded cables.

10.6 B.S. Tests for Mains Transients in Electronic Control Equipment

A range of tests is defined in BSS 4509, 1970, and BSS 4525, 1970, in which short-duration interruptions (5, 20, 100 and 500 ms) are introduced into the mains supply and also peak overvoltages randomly phased with respect to the supply (see also EES, 1970) (CEGB[d]). The peak half sine form overvoltages, expressed as a percentage of the mains rms voltage, are:

Peak Overvoltage, %	Pulse Duration
10	10 ms
200	1 ms
300	20 μs
500	5 μs

The equipment, designed by the Scientific Instrument Research Association, is manufactured under licence (Entry 43, Appendix A). The instrument is made in two types as follows:

(a) E55005 generates transients having a pulse duration of between 5 μs and 1 ms.

(b) E55006 generates transients having a 10 ms duration and provides mains interruptions of 5, 20, 100 and 500 ms.

It should be noted that the risetime of the transients is unspecified and that owing to the relatively long duration of even the shortest pulse (5 μs-200 kHz) the test is unlikely to reveal susceptibility to very short interference pulses a few tens of nanoseconds (ns) long, which it has been shown (Chapter 7) are those to which integrated-circuit systems are most susceptible.

10.7 Beckman Line Noise Generator

This commercial instrument of U.S. origin (Entry 8, Appendix A) generates controlled mains supply noise up to 10 kW peak power, the noise pulse amplitude being variable from 0 to 600 volts. The pulse risetimes are better than 0.5 μs; the repetition rate is at half the supply frequency, the pulse being capable of manual phase adjustment through 0-360 degrees or capable of being swept through the range in 3 seconds.

10.8 Shaffner Equipment

The equipment NSG 200C, of Swiss manufacture, comprises a main frame and plug-in drawers containing mains interrupters and alternative interference sources (Entries 39, 72, Appendix A). The main frame NSG 200C contains line switches and filters and accommodates any one of four (at present) alternative plug-in units.

NSG 202 simulates interruptions of line voltage
NSG 203 simulates short variations in line voltage
NSG 222 simulates short pulses with fast risetime
NSG 223 simulates 50 microsecond interference pulses with high energy.

Additional plug-in units for tests according to MIL and NATO specifications are in preparation.

Some abridged details of the four units are given below:

(a) Plug-in unit type NSG 202

Interference corresponds to mains interruptions of BSS 4509 and provides for total interruption of line power for periods of 1.5 ms to 500 ms, repeatable at intervals of 0.6 s to 25 s. The maximum current capability is 16 A at 50 Hz, single phase. The test is intended to reveal loss of regulation or loss of information caused by line interruptions.

(b) Plug-in unit type NSG 203

The interference is generally similar to that of (a) above, but the interruptions need not be total. The line voltage may be temporarily reduced to a preset lower value, thus simulating the effect of short (0.02 to 5 s) variations in line voltage such as those occurring during system faults and having durations of 0.02 to 5 s. The current capability is 15 A at 50 Hz, single phase. The test is intended to reveal equipment malfunctions, possibly destructive.

(c) Plug-in unit type NSG 222

The interference comprises fast low-energy pulses with a risetime of 5 to 30 ns, an amplitude of 500 V to 1.5 kV and a duration of 0.1 microsecond. The interference is intended to simulate the operation of electromechanical switches, relays, etc., and is thus capable of revealing malfunctioning of I.C. logic circuits. Since the pulses have a very fast risetime, a broad frequency spectrum is produced. A capacitor is charged to 600 volts and periodically discharged through a thyristor into the primary of a transformer having a ratio of 1:12. The transformer output charges a capacitor which may be discharged through a spark gap. The pulse height is 500 V, 1 kV or 1.5 kV, from an impedance of 150 ohm in series with a 5 nF capacitor. The resulting pulse can be injected into the mains (common mode) or coupled to a data-transmission line. Alternatively, a static discharge may be simulated (6 kV, 150 pF, 20 kΩ in series, corresponding to a person in a dry atmosphere touching an earthed surface). The operation may be continuous, single shot or gated with adjustable phase from the mains.

(d) Plug-in unit type NSG 223

The unit superimposes potentially destructive pulses on the mains voltage or on other supply voltages, including D.C. supplies, in either series or common mode form. The pulse output is available at 1, 3 and 5 kV with internal impedances of 5, 50 and 125 ohm, respectively, and may be continuous, single shot or gated with adjustable phase. The risetime is 0.3 microseconds of either polarity. The principal application lies in the testing of components.

(e) Unit type NSG 504

There is also a self-contained Schaffner impulse-voltage generator NSG 504 corresponding to IEC Publication 60 and the recommendations of IEC Technical Committee 41 (see also Section 10.1). The NSG 504 is intended to test mechanical and static relays, cables, counters, switches, resistors, capacitors, transformers and chokes. The pulses have a risetime of 1.2 μs and a length of 50 μs with a maximum amplitude of 5.5 kV. The internal impedance is 500 Ω. The pulse energy may be adjusted from 0.5 to 2.5 Joule, and operation may be continuous at a repetition rate of 0.1 to 1 Hz or single shot.

10.9 Power Line Disturbance Monitoring

Clearly the definition of mains transient withstand levels for any particular installation, or type of installation, will be more meaningful if the extent of knowledge of the magnitude of actual transients (Turner, 1968) is increased. This involves the continuous monitoring of actual supplies in a real installation; instruments such as those used by Turner (C.E.R.L. Surge Monitor) may be used, or other special-purpose sophisticated disturbance monitors. The MPL monitor, for example, counts and records the time of incidents such as excess frequency error, rms over- and under-voltage and bipolar transients of 0.2 μs duration or more exceeding two preset values (Entry 40, Appendix A). It should be noted that simple mains check sets are less suitable in that in most versions the response to transients falls off rapidly with pulses of 1 μs duration, and it has already been shown that we are concerned with pulses of even shorter duration.

10.10 Electromagnetic Field Withstand Test, Laborelec

The computer, or other vulnerable equipment, is set up to operate normally on a test program and is completely enclosed in a solenoid winding energised from the appropriate source. The winding is placed in each of the three mutually perpendicular planes, and currents as follows are each applied at least three times:

(i) 50 Hz

(ii) Damped train of sine waves having a fundamental frequency which is adjustable and extending to several MHz. In the tests so far done, the particular interference source circuit available limited the upper frequency to about 4 MHz (see Chapter 6); although ideally, because of the increasing sensitivity with frequency, fields at frequencies extending to 30 or more MHz are desirable.

Laborelec consider that for classification (a) (see Section 10.4) the equipment should remain unaffected by fields of 50 A peak/m, whereas in classification (b) a lower value of 10 A peak/m should be acceptable.

The results quoted in Chapter 6 were obtained with the method outlined.

It should be noted that the UKAEA method (Section 10.11) is capable of being used in a similar manner, the vulnerable equipment being exposed to induction from pulse and radiofrequency currents between 10 kHz and 100 MHz flowing in a suitable injection line, the criterion being that operation is unaffected, or, where appropriate, that the measured disturbance is below a specified value.

10.11 UKAEA Interference Immunity Tests

The UKAEA method is considered to be the most valuable means of testing and quantifying the interference immunity of individual items of a complete installation and also of the installation as a whole. Because the noise environment can change with time, there is a strong reason for using an arbitrary test level to assess the necessary improvement in any situation. Fowler (1974) has very clearly defined the method in AEE Winfrith Report R919. Although the objective has been to secure immunity of highly sensitive wideband nucleonic instruments in the nuclear-power-station environment, the considerations are basic and are fully applicable to the present case.

Fowler and Gregory (1970) refer to the existence of earth currents of up to 500 A at 200 kHz in earth connections near the gas-circulator motors at Wylfa, and Report R919 quotes 100 mA rms trnasient earth currents with frequency components extending to beyond 100 MHz in cable sheaths, due to the switching of supplies at only 3.3 and 11 kV. The objective of the work has been to specify the acceptable levels of common and series mode interference over a frequency range extending to 100 MHz for various nucleonic instruments. Fowler has pointed out that interference immunity is often degraded by nonlinearity in the active circuits and that, because of this, consideration must be given to interference including frequencies *outside the normal passband of the instruments.*

The principle of the UKAEA tests (IEE, 1970) is based on the injection of a suitable test signal into individual units including input and output connectors whilst observing the effect on the vulnerable equipment. Imperfections in the screening result in a component of the transfer impedance derived from asymmetric penetration of the screen known as shine. The impedance Z_{sh} is the ratio of the voltage induced in the screened circuit due to this effect to the interfering current in the external circuit. Z_{sh} varies with the distance of the current path from the screened circuit and for connectors is measured with a distance of 10 mm. Further details are given by Fowler in Appendix B.

In the first instance, the signal source comprises a power amplifier capable of delivering 100 mA rms into a short circuit (or a voltage of 10 V rms into an open circuit) and covering the whole frequency range of 10 kHz to 100 MHz. When shine tests are made, a 200 ohm impedance balanced transmission line is employed as a radiator. This source is used to explore the screening of an instrument or connector. The line is driven at 100 mA rms over the whole frequency range.

Injection tests are made on individual units by direct connection of the 100 mA source into the instrument chassis, between various points on the earthed chassis, from the cable screens adjacent to each connector to the earth point, etc.

Because the mains-supply impedance is about 50 ohms at frequencies exceeding 100 kHz and disturbances can reach at least 340 volts peak amplitude, an immunity to 8 amperes in a properly matched system is appropriate. Owing to mismatch at filters, voltages exceeding 1 kV peak-to-peak can arise. The UKAEA proposal for a test signal for mainsborne interference is a low-repetition-rate pulse having a fast risetime of 10 ns and 1 microsecond duration. (It should be noted that by contrast the risetime of the pulse in the BS instrument [Rosemount] does not appear to be specified; that of the Beckman instrument is 0.5 μs. Consequently, resonances in the higher-frequency range which could be detrimental to vulnerable equipment may not be so easily revealed by these devices.) The pulse amplitude is 400 volts across a 50 ohm load impedance. The pulse signal is applied via capacitors in both series and common mode connection to the mains supply. A power pulse amplifier (or an alternative high power pulse generator) is required.

When the individual components have been proved immune to the requisite level, the complete assembly including interconnecting cables can be tested using the same source feeding into an injection line. This line comprises a 2.5 mm^2 p.v.c. insulated single wire laid roughly within 1 m (and where possible spaced 50-200 mm from conducting metal obstacles) of the instrument system signal cables whose screening is under investigation. The signal injection line must be run beyond the far end of the instrument cables and is terminated on a noninductive resistor of 180 to 220 Ω connected to the local station earth at the far end. The 100 mA (or more using the power pulse generator) test signals are injected into the line, and the coupled interference is displayed and measured on a fast oscilloscope joined to the equipment undergoing test.

The UKAEA have suggested that, for certain instruments, allowable levels of disturbance caused by 100 mA rms flowing in any path of an instrument or in a nearby conductor (shine tests) over the frequency range 10 kHz to 100 MHz should be:

Linear indicators	5% FSD
Trip circuits	10% margin to trip
Nonlinear mean DC measuring circuits	2 times standard deviation from circuit noise or signal characteristics

An Interference Test Unit (95/4201-3/4) has been designed by UKAEA to provide the requisite signals and can be obtained from a manufacturer (Entry 17, Appendix A). The unit generates alternately two pulses and two bursts of radiofrequency, the frequency of the latter being continuously adjustable (to investigate resonance effects) between 10 kHz and 100 MHz. The details are given in the UKAEA report (Fowler, 1974). The test unit is intended to be used with Interface Unit 94/5238-1/4 and a Hewlett Packard type 180A oscilloscope. Alternatively, Interface units 94/5239-1/4 and 94/5240-1/4 can be used with Tektronix type 454 or 7000 series oscilloscopes, respectively. Particulars of the Interference Test Unit and interfaces are given in the relevant technical manual published by the manufacturer (Entry 17, Appendix A).

In data-processing and computer systems, higher-power test signals than can be achieved by the Interference Test Unit unaided are required. A portable 5 kW pulse generator Type 95/4221-1/4 has therefore been developed (see Fowler, 1974, Appendix II). The Pulse Generator is made commercially (Entry 17, Appendix A), and the technical manual published by the maker gives full details. The pulse generator provides a 10 A pulse into a 50 Ω load (5 A into 200 Ω) with a fast risetime of 10 ns and a duration of 1 μs. The repetition rate may be set, the maximum being 1 kHz.

Fowler shows that a *5-10 A peak amplitude pulse may be necessary to achieve similar immunity in a computer* to the accepted 100 mA rms burst from Interference Test Unit 95/4201-3/4 in sensitive analog equipment. The method of use of the 5 kW pulse generator (which is portable and battery powered) is similar to that of the Interference Test Unit. The pulses are delivered into an injection line terminated in 200 Ω, and the effect on the vulnerable circuit or equipment is observed on an oscilloscope. The pulse current is displayed on one beam and the effect on another beam. The pulse amplitude is raised until faulty operation is apparent. If no faulty operation is observed at full amplitude, the system is given a clean bill of health at that amplitude (read from the controls or from the oscilloscope).

An accessory box can be used to allow pulses to be injected into the mains supply.

The principal disadvantage of this method, unlike the former, is that resonances, although apparent, cannot readily be explored by sweeping the frequency.

It is appropriate therefore to investigate every circuit with the Pulse Generator 95/4221-1/4 and to use the Interference Test Unit 95/4201-3/4 as a diagnostic tool in making further tests on any circuits shown up by the pulse-generator test as being in need of improved filtering or screening.

11 Conclusions

The conclusions are numbered to correspond with the Recommendations (Chapter 12).

11.1 Operation of Electronic Systems in Substations

Providing known precautionary measures are taken, there will be few problems in the operation of digital or integrated-circuit equipment in high-voltage stations. However, quantitative data on individual plant susceptibility and on interference levels and spectra need some reinforcement before measures can be concisely defined for each and every case. For the reasons given in Section 11.2, precise definition is, in a sense, unnecessary.

11.2 Extent of Measures Needed in Future

Because of the high and increasing rate of development and consequently of obsolescence and irreplaceability of electronic components, there is a risk that, if only the minimum measures are taken to permit the use of a present-day system, then replacement of components or the system itself at some future time may result in interference problems again becoming manifest. This is so because integrated-circuit designers are currently providing up to 10,000 devices in one chip and are moving towards incorporating 250,000 elements in the one chip, perhaps rendering it more susceptible to interference and damage.

Integrated circuits reach a maximum interference sensitivity at frequencies of tens of MHz, and computers are currently using signals a few tens of nanoseconds in duration. Consequently, interference suppression at frequencies extending to 100 MHz is already a requirement. Adequate screening and filtering will be essential. It seems likely that a screening and filtering attenuation of not less than 60 dB over the frequency range extending to 100 MHz should be adequate as an aim, but this remains to be proved; see Chapter 8. The introduction of optical-fibre systems could well avoid all need for measures other than the electromagnetic screening of vulnerable equipment. *It would be prudent to take all practicable electromagnetic-compatibility measures in the long-life permanent installations even though they may not all be strictly essential at present.*

11.3 Shortcomings of Present Practice

Inspection of installations designed for conventional substantially nonelectronic systems in power stations and high-voltage substations confirms that, in relation to the aspects considered in this book, the current installation practices are sometimes not ideal for vulnerable electronic systems.

It is clearly necessary to achieve single-point earthing of instrumentation circuits (from interference considerations). At the same time, the requirement to bond and earth accessible metalwork including cabinets (for touch potential safety reasons) is to a degree in conflict with this requirement. The solutions foreseen lie in the use of barrier devices to permit each item of equipment to be referred to its local earth and, where this is not possible, to use double enclosure; the inner screen floating in potential with the source and the outer screen bonded to local earth.

11.4 Earthing Practices

Because of the high frequencies which can be damaging and the significant impedances presented by even a few feet of conductor, however heavy, reliance on conventional earthing practices alone is not sufficient. Furthermore, the existing earthing practices could readily be improved to reduce interference radiation and conduction.

11.5 Screening

The existing screening is often inadequate or absent. Resort must be made to improved magnetic and radiofrequency screening by enclosure of all vulnerable equipment. Conventional apparatus cubicles provide poor radiofrequency screening and could readily be enhanced (Figs. 6 and 41).

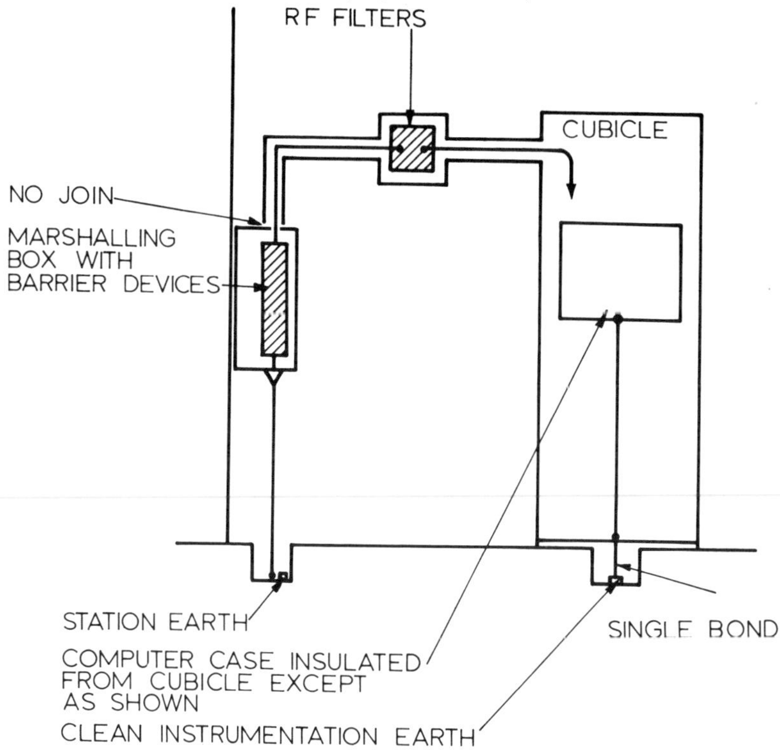

G.41 PRACTICAL INSTALLATION SHOWING BARRIER DEVICES, FILTERS AND EARTHING ARRANGEMENT

11.6 Filtering

Because metallic connections are made to present-day vulnerable equipment, these must be filtered at the point where they pass through the screening enclosure, as otherwise the value of the screen is reduced. Such filters are compact and inexpensive. Attenuations of 40-60 dB will be required in the frequency range 100 kHz to 100 MHz, depending on the vulnerability of the particular equipment and the interference exposure.

11.7 Barrier Devices

Common mode (rise of earth potential) interference is the most prevalent problem in high-voltage substations, whereas both common and series mode can be serious in generating stations, where larger numbers of low-level analog signals are handled. Barrier devices will continue to be a necessity to counter these effects. A variety of barrier devices (relays, optocouplers, transformers, amplifiers, etc.) exist, but attention needs to be given to their choice and to improving their properties at higher frequencies.

11.8 Standard Interface

It would be advantageous to devise a standard interface unit (containing barriers, limiters, attenuators and filters) suitable for the various common types of signal. Such an interface would achieve flexibility of choice of electronic system as well as facilitating changes to the latter.

11.9 Instrument Transformer Earthing

The circuit practice used conventionally on some instrument transformers, whereby the secondaries are earthed at a point distant from the source, enhances the chance of interference by providing a large area inductive loop and encourages rises of earth potential. Twin twisted balanced about earth circuits would be advantageous. CT, VT and carrier coupler circuits provide a path for interference into the apparatus room, and the application of suitable earthing, signal transmission and filtering measures is clearly essential.

11.10 Inadequate Radiofrequency Screening and Filtering in Commercial Equipment

Because many production equipments are designed for use in essentially nonhostile environments (and owing to commercial considerations will perforce have to be utilised in the substation environment), it should be realised that their screening, filtering and internal design practices are often such as to render the equipment more vulnerable to interference than is desirable, thus emphasising the need for additional screening and filtering precautions to be added.

11.11 Design of New Equipment

Where new equipment designs are being made, the opportunity exists to incorporate appropriate measures to achieve interference immunity by choice of philosophy and technology used in the circuits as well as by incorporation of appropriate screening and filtering.

11.12 Conducted Interference Levels

There are some published data on the amplitudes of conducted interference impressed on the small wiring of substations, but generally the results refer only to peak amplitudes and do not give the spectra. As may be expected, current transformer, voltage transformer and power line carrier coupler circuits are the worst offenders, giving voltages of several kilovolts, particularly during switching and isolator operations. A withstand level of 8 A using the UKAEA test appears appropriate for power-supply connections.

11.13 Radiated Interference Levels

There is very little published information on electromagnetic field levels or spectra existing for short periods during switchgear and isolator operation, or as a result of 50 Hz system faults. However, the few figures given in this book could be used, supported where necessary by calculations based on simplified assumptions of circuit constants.

The normal radio interference measurement methods (BS 727:1967) used for studies of corona noise are not convenient for the examination of the spectra of short-duration bursts of electromagnetic radiation.

11.14 Withstand Levels, Conducted Interference

The acceptable withstand levels depend on the individual devices being considered and the manner in which their function is or could be degraded. They are generally not known at present. Each device and system requires separate consideration until such time as a catalogue of acceptable interference level limits for commonly used devices and instruments can be built up. Generally speaking, it should not be difficult to achieve attenuations erring on the safe side, so that an order of magnitude rather than a precise definition may well be adequate; reliance being placed on barriers, attenuators and filters (see also Chapter 8).

11.15 Withstand Levels, Magnetic Interference

Until the vulnerability of different items of plant is known by test, the indications are that all vulnerable equipment should be called for to operate without degradation in a field (in any plane) of, say, 100 A_c/m in the frequency range 50 Hz to 100 MHz.

11.16 Test Methods and Specifications

It appears that there are at present no interference withstand levels specified for each item of substation or power-station light-current plant, nor for that matter for complete systems. Measures to combat interference require the equipment designers to state the maximum allowable levels of both conducted, inductive, and radiated electromagnetic interference in which each item of equipment can still operate correctly within its specification.

Such figures will include:

(a) The acceptable series mode interference levels in the normal passband for conducted interference.
(b) The acceptable series mode interference levels over the frequency range from D.C. to 100 MHz for conducted interference.
(c) The acceptable common mode interference levels over the frequency range from D.C. to 100 MHz for conducted interference.
(d) The maximum tolerable magnetic and radiated electromagnetic fields in the frequency range 50 Hz to 100 MHz (in any plane) at each type of vulnerable equipment.

It should be particularly noted that, in a given instrument, because of nonlinear effects, interference well outside the passband required for the measurand may still be troublesome.

The interference withstand levels cannot be determined by the use of the usual BEAMA or other voltage withstand tests alone. Such tests will therefore need to be supplemented by a combination of the tests outlined in Sections 10.8 (Schaffner), 10.10 (Laborelec) and 10.11 (UKAEA) and in Appendix B (UKAEA). These tests are repeatable and offer distinct advantages.

11.17 Specification of Barriers, Filters and Screening

As experience is obtained in the susceptibility of the various vulnerable plant items and as gaps in the knowledge of the environmental conditions in substations are filled in, a catalogue of interference susceptibility and interference environment values will be built up (Chapter 8 refers). This can then be used to specify the attenuation values actually required. Present indications are that attenuation will be required for both conducted and radiated interference at the point of ingress into the screened enclosure and in the frequency range 10 kHz to

100MHz, although actual values must depend on the individual equipment and its environment.

11.18 Action to Check on Current and Future Installations

There is at present no central register of interference levels or plant withstand values. Similarly, the growing use of electronic plant makes a common clearing point for interference matters essential in the specification, installation and retrospective fitting stages, so that all information is pooled and standard methods of test are brought into use.

11.19 Introduction of UKAEA and Schaffner Tests

The UKAEA test method (Fowler, 1970), in which short high-powered pulses and bursts of radiofrequency energy are injected into an individual device or complete system, permits the interference immunity of individual items of plant and of complete installations to be measured quantitatively (Entry 17, Appendix A). This will require an injection line to be laid parallel to the vulnerable cable routes. The introduction of the UKAEA test method is strongly recommended.

The tests made possible by the Schaffner equipment (Section 10.8) are more comprehensive than others, and use of this equipment in addition to the UKAEA tests would be beneficial in quantification of interference immunity levels.

11.20 Optical Systems

Optical communication systems are feasible over distances of the sort required for the replacement of substation multicore cables, and commercial components are increasingly available. Use of optical fibres (as opposed to optocouplers) or free-space systems can remove the conducted interference problems (Section 4.5.9), leaving only the screening requirement for electrical/electronic apparatus.

12 Recommendations

The recommendations are numbered to correspond with the Conclusions (Chapter 11).

12.1 Operation of Electronic Systems in Substations

Where the costs are justified, the improved screening, signal transmission, earthing, barrier and filtering measures outlined in this book should be employed in all vulnerable installations, with the aim of achieving not less than 60 dB screening and filtering attenuation of the vulnerable equipment. This can readily be obtained by inexpensive measures.

12.2 Extent of Measures Needed in Future

All practicable electromagnetic-compatibility measures should be taken in the long-life permanent installations even if not strictly essential at present.

12.3 Shortcomings of Present Practice

Existing installations should be modified having regard to the economics so as to include the enhanced screening, earthing, barrier and filter methods described.

12.4 Earthing Practices

Where it is proposed to install vulnerable equipment, a careful study should be made of all earthing arrangements in the light of the information given.

12.5 Screening

Attention should be given to the closer specification of radiofrequency screening enclosures to surround every vulnerable item. The cases of many instruments could provide sufficient screening if appropriately modified. The possibility of providing a modification to the standard cubicles so as to obtain more effective screening by the use of finger contact strips and closure of holes (including the base) should be considered.

12.6 Filtering

The point of entry of each and every conductor entering the screened enclosure wherein the computer or other vulnerable equipment is housed should be provided with a radiofrequency filter giving at least 40 dB attenuation over the frequency range 1 MHz to 100 MHz. As the present sensitivities of vulnerable equipment are unknown, higher filter attenuations of 60 dB may be a necessity in some cases. Such filters are readily obtainable, compact and inexpensive. Information on makers is given in Appendix A (Entries 9, 51, 58, 62, 68, 71, 72).

12.7 Barrier Devices

It is recommended that a standard range of barrier devices having known properties should be evolved, to include modified versions of existing reed interposing relay units, optocouplers, transformers, amplifiers, etc. In all cases the high-frequency performance needs to be improved by attention to installation details and by incorporation of Faraday screens. The interference attenuation properties should be established of a range of standard types of barrier and should be catalogued. No new device should be brought into use without knowledge of its properties over the whole frequency range of interest.

12.8 Standard Interface

Consideration should be given to incorporating the barrier devices, limiters, attenuators and filters in a modular form into a standard interface of known properties, so that any properly screened electronic system could be attached with full confidence that interference levels would be held to defined values.

12.9 Instrument Transformer Earthing

Traditional instrument transformer earthing practices are conducive to the enhancement of common and series mode interference and, in some cases, to its propagation into the apparatus room (Chapter 5). Consideration should be given to using twin twisted and screened cables and balanced signal transmission by means of interposing balun transformers, retaining the instrument transformer secondary earth near to the h.v. source. Where only relatively narrow bandwidth signals are required, as is often the case, filters incorporating series attenuation and shunt paths to bypass interference should be considered for installation close to the source, near to the instrument transformers.

12.10 Inadequate Radiofrequency Screening and Filtering in Commercial Equipment

Where production equipments designed for essentially nonhostile environments are chosen for use in high-voltage stations, care should be taken to select equipment incorporating the best screening and incorporating internal design practices in accordance with this book so as to use equipment having the maximum immunity to both radiated and conducted interference.

12.11 Design of New Equipment

Where electronic systems are being developed for use in the

hostile environment of a high-voltage station, attention should be given to the specification of conducted, inductive and radiated interference withstand levels and to the choice of system philosophy and technology (choice of i.c. type, etc.) to obtain adequate interference immunity. Proper earthing, filtering and screening precautions, in accordance with this book, should be built in and should not require to be applied retrospectively.

12.12 Conducted Interference Levels

Steps should be taken to supplement the available data on conducted interference amplitudes during switch and isolator operation by some measurements of the spectrum of bursts of such interference, particularly on current and voltage transformer circuits and on carrier coupler circuits, both at the source and in the apparatus room, for each system voltage. This may require instrument development in common with Section 12.13.

12.13 Radiated Interference Levels

The knowledge on radiated interference in the vicinity of high-voltage plant is chiefly limited to measurements of corona noise. Since these are unsuitable for the purpose, some measurements should be made to supplement the information given in this book. The measurements should be aimed at discovering the field and its spectrum during switching operations.

12.14 Withstand Levels, Conducted Interference

12.15 Withstand Levels, Radiated Interference

Because the values are largely unknown, urgent steps should be taken to build up a catalogue of the maximum tolerable values of both radiated and conducted electromagnetic interference which is allowable for each item of vulnerable plant, to investigate the susceptibility of new items and to maintain an up-to-date record, enabling relative susceptibilities to be quantitatively compared and appropriate suppression and plant choices to be made. The use of the Schaffner instrument (Section 10.8) and UKAEA shine test methods (Section 10.11) (Fowler, 1974) or similar is strongly recommended, together with the further development of the Laborelec electromagnetic inductive field exposure test (Section 10.9). It is recommended that such tests should cover a frequency range as wide as possible (up to 100 MHz in the UKAEA test), since a 1/50 μs test cannot alone disclose screening and filtering deficiencies.

12.16 Test Methods and Specifications

Interference withstand, screening and filtering tests should be made on existing plant and new designs of plant at the prototype stage. Complete installations including the final cabling should be tested by the UKAEA method. Injection lines for this method should be provided along the routes taken by the cables of vulnerable circuits, so that periodic checks can be made to disclose any deterioration of screening attenuation.

Since there are at present no interference withstand levels specified for each item of plant or for complete systems, the equipment designers should be required to state the maximum tolerable levels of radiated, inductive and conducted interference in which each particular equipment can still operate correctly within its specification, using the test methods defined (Section 11.16). These values will need to be related to the interference environment (Section 12.12, 12.13) before the required attenuation (Section 12.17) can be specified precisely.

12.17 Specification of Barriers, Filters and Screening

When sufficient information of the interference levels (Sections 12.12, 12.13) and of the withstand values of each type of vulnerable plant (Section 12.16) is provided, the required signal to interference ratios for common and series mode interference and radiated interference enable the attenuation to be specified.

12.18 Action to Check on Current and Future Installations

Insofar as the electricity supply utility industry is concerned, it is suggeested that a method should be established whereby each installation is scrutinised at the planning stage for compliance wherever possible with the contents of this book. The information so obtained should be used to provide formal specifications for the application of the defined test methods and for installations themselves. Such scrutiny could advantageously be applied to existing installations where new vulnerable equipment is to be interfaced.

12.19 Introduction of UKAEA and Schaffner Tests

Conventional voltage withstand tests on vulnerable equipment should be supplemented, on individual items and on whole systems, by the UKAEA interference or similar wideband immunity tests, using the 5 kV pulse generator (Fowler, 1970; Entry 17, Appendix A). An immunity level of 8 A injected into supply circuits is appropriate. The immunity level to be fixed for signal circuits will depend on the outcome of Section 12.15.

12.20 Optical Systems

Optical-fibre and free-space optical communication systems are available and are becoming increasingly competitive for distances of a few hundred meters. The cable itself is largely immune to interference problems, so the opportunity should be taken to effect an installation in a substation where vulnerable equipment would have all its input and output signals coupled via fibre optics. The costs of such a step should be compared with the costs of undertaking the measures recommended in this manual for interference reduction in conventional metallic wire systems.

13 Acknowledgments

The author acknowledges the assistance of Mr. F. Cross, Mr. L. Csuros and Mr. J. Eastaugh of the CEGB Transmission Development and Construction Division in providing material and information, and for their many helpful comments.

The help of the many other colleagues who provided information, and particularly H. McFarlane (CERL) and Mr. J.C. Metcalfe (MEL), is acknowledged.

The permission of Laborelec (Bruxelles) to abstract information including Table 3 from a number of their reports is gratefully acknowledged, and thanks are also due to Mr. R. Wery and his staff at Laborelec for their help in extensive discussions.

The permission of CIGRE and of the authors to reproduce in Table 2 and Figure 3 material from the CIGRE paper by Anders and Campling, 36-09, 1976, and in Figure 39 material from the CIGRE paper by Borgvall, Holmgren, *et al.*, 36-05, 1970, is gratefully acknowledged.

The help of Mr. E.P. Fowler (UKAEA) for his valuable comments and especially in his making available all the material incorporated in Appendix B is also warmly acknowledged.

This book is published by permission of the Central Electricity Generating Board, London, England.

14 References

Allan, A., "Noise immunity comparison of CMOS versus popular bipolar logic families," Motorola application note AN-707.

Anders, R., and Andersson, J., "Optronic protection system for series capacitor platforms," CIGRE Paper 36-77 (WG04) 06E, 1977.

Anders, R. and Campling, A.C., "Investigations into interference in substation and power station auxiliary cabling," CIGRE Paper 36 09, 1976.

Barratt, P.J.S.B., "An investigation of the field disturbance caused by operating a 275 kV air blast circuit breaker at Tilbury sub-station and its effect on the recording equipment of the MDRU." Unpublished CERL internal note, September 1965.

Birch, F.H., Burrows, G.H., and Turner, H.J., "Investigations into transient overvoltages on secondary wiring at EHV switching stations," CIGRE Paper 31-04, Part II, June 1968.

Birch, F.H., and Turner, H.J., "Field investigations into transient overvoltages in secondary wiring." IEE Colloquium Digest No. 1968/19, pp. 382-390, 1968.

Bishop, A., "Comparison of low power Schottky T.T.L. and C.M.O.S.," *New Electronics,* 15 October 1974.

Blood, B., "A.C. noise immunity of MECL 10,000 integrated circuits," Motorola application note AN-592.

Borer, "Input-output adapters for CAMAC. Types 1941, 1942, 1943." Borer ref. 603.3.059.3.74. Borer Electronics A.G., Solothurn, Switzerland.

Borgvall, T., Holmgren, B., Sunden, D., Widström, T. and Norback, K., "Voltages in substation control cables during switching operations," CIGRE Paper 36.05, 1970.

Brace, D.J., and Ravencroft, I.A., "Optical fibre transmission systems: the 8,448 Mbit/s feasibility trial," *P.O.E.E.J.,* Vol. 70, Part 3, October 1977.

British Electrical and Allied Equipment Manufacturers Association, "Recommended transient voltage test applicable to transistorised relays," Publication 219, 1966.

Brown Boveri Review, Vol. 61, No. 8, August 1974.

Bull, J.H., "Electrical interference with electronic equipment," Electrical Research Association, ERA 71-50, April 1971.

Bull, J.H., "Code of practice for the avoidance of electrical interference in electronic instrumentation and systems," Electrical Research Association, ERA 75-31, March 1975.

Campling, A.C., "Interference voltages and screening of auxiliary cables in substation switchyards," CIGRE Paper SC 36-77 (WG04) 02E, 1977.

C.C.I.T.T., "Directives concerning the protection of telecommunication lines against the harmful effects from electricity lines," I.T.U., Geneva, 1973.

CEGB(a), "General Specification for Reed Relays," (RR 1972).

CEGB(b), "Specification for protection equipment to protection signalling/intertripping equipment interface with 10 kV isolation barrier interface unit type II" (PSI 876 AG) (CEGB P12-1974), 1974.

CEGB(c), "Interfacing of telecommunication equipment with other users at outstations," Design Guide No. 4, Issue 3, 12/12/73.

CEGB(d), "General specification for electronic equipment," Specification CEGB-EES 1970.

Cerberus Bulletin No. 38, September 1975, Cerberus A.G., Mannedorf, CH 8708, Switzerland.

C.I.G.R.E., "Teleprotection," 1969, C.I.G.R.E., 112 Bvd. Haussmann, 75008, Paris.

Clare, 1975, Technical Application Reference, Mercury Wetted Relays." C.P. Clare Corp., January 1975.

Conductive Concrete: "Marconi Communication Systems Ltd. Conducting Concrete: Marconite," *Electronics Industry*, January 1978, p. 47.

Cooke-Yarborough, E.H., Franklin, E., Grisow, J., Mowlett, R., and West, C.D., (a) "Thermomechanical generator: An efficient means of converting heat to electricity at low power levels," *Proc. IEE,* Vol. 121 (7), pp. 749-751, 1974.
(b) "The Harwell Thermo-mechanical Generator," UKAERE Report R7714, 1974.

Cox, R.J., "Electronics and Nuclear Power," *The Radio and Electronic Engineer,* Vol. 45, No. 10, pp. 603-609, October 1975.

Cree, D.J., and Giles, L.J., "Practical performance of radiating cables," *The Radio and Electronic Engineer,* Vol. 45, No. 5, pp. 215-223, May 1975.

Cushing, V.J., "Transient suppression in 500 kV installations," IEEE Conference Paper C74 030-3. IEEE PES winter meeting, New York, January-February 1974.

Dixon, J.F., "Report on microprocessors/microcomputers," ref. DT/JFD/573.10.430, April 1975, Reprinted April 1976, Transmission Division CEGB.

EES 1970, See General Specification for Electronic Equipment.

Electrical Research Association, "Interference with electronic instrumentation and systems," Reference 3041/68.

Feist, K.-H., "Earthing problems in SF_6 stations." CIGRE Paper 36-77 (WG04) 09E, 1977.

Ficchi, "Electrical Interference," Iliffe, 1964.

Fowler, E.P., "Some recent work on the screening efficiency of braided and multiscreened signal cables," Conference on Electrical Interference in Instrumentation, IEE Conference Publication No. 65, pp. 23-33, June 1970.

Fowler, E.P., Interference Immunity Tests for Nucleonic Instrumentation," U.K. Atomic Energy Authority Report AEEW R919, April 1974, HMSO.

Fowler, E.P., and Gregory, H., "Studies of the source of electrical transients and interference at Wylfa Nuclear Power Station," IEE Conference Publication No. 65, pp. 81-99, June 1970.

"General Specification for electronic equipment," Specification CEGB-ESS, 1970.

Harrison, D., "The mechanism of interference pick-up in cables and electronic equipment with special reference to nuclear power stations," *The Radio and Electronic Engineer,* Vol. 29, No. 3, pp. 149-58, 1965.

Harrison, D., Wilson, I., and Cox, R.J., "Reducing the effects of electrical interference in nuclear instruments," Proceedings IAEA Nuclear Electronics Conference, Bombay, November 1975.

Harvey, R.P., "The scanning of low level signals for computer inputs." IEE Conference Publication No. 65, pp. 50-53, 1970.

Harvey, S.M., "Ontario Hydro transient immunity test." CIGRE Paper SC 35/36, Toronto, 1977.

Healey, A.M., Mrs., "Bibliography of electrical interference in Instrument and Control Engineering," Corporate Development Laboratory, British Steel Corporation, 1973. (Covers 1965-72 and *includes 297 selected references* relevant to instrumentation and control engineering in the process industries. Aspects included are sources and occurrence; effects on systems and instruments; measurement and prediction; and techniques for the control of interference. The fields of radio and communications engineering are excluded.)

Hewlett Packard, "High speed optically coupled isolators." Application Notes: 939, "Balanced line receivers";
948, "Performance in short/medium length data systems";
951-1, "Applications for low input current high gain optically coupled isolators."

Hooper, J., "Connection of L.V. power supplies to signalling equipment on high voltage transmission towers," CERL Note No. RD/L/N 165/73.

IEE, "Electrical Interference in Instrumentation," Conference Publication No. 65, 1970.

IEEE, "The Neutralising Transformer Concept for Protection of Wire Line Facilities entering Electric Power Stations," IEEE Interim Committee Report F76, 379-8. PES Summer meeting, Portland, Oregon, July 18-23, 1976.

Jackson, I, "Practical braid-breakers using stock materials," *Radio Communication,* November 1972.

Jacquet, B., "Problèmes de pertubations induites dans la filerie auxiliaire des postes à haute tension, Part II." Laborelec, 1977.

James, K.W., Kingham, E.G., and Moore, D.J., "Data acquisition systems for measuring atmospheric pollution," IMEKO VI, Section 6, 1973.

Jones, F., "The shielding factor of underground cable sheaths," IEE Monograph No. 59, January 1953, *Proc. IEE,* Vol. 100, IV, No. 1, pp. 34-38; also *Electronics and Power,* Vol. 20, No. 21, p. 1051, November 1974.

Klewe, H.R.J., "Interference between power systems and telecommunication lines," Edward Arnold, 1958 (also ERA report M/T 126).

Knox, K.A.T., "Semiconductor devices in hostile electrical environments," *Electronics and Power,* Vol. 19, pp. 557-560, December 1973.

Kunkel, G.M., (a) "Practical and theoretical aspects of single point grounding," IEEE International Symposium on Electromagnetic Compatibility, 1970.

Kunkel, G.M., (b) "Induced voltage in system contents due to the complex importance of ground reference lines," IEEE Electromagnetic Compatibility Symposium Record, 1970.

Malmström, U., "Disturbances in computer systems." CIGRE Paper 36-77 (WG04) 07E, 1977.

Martin, D.J.R., "A general study of the leaky-feeder principle," *The Radio and Electronic Engineer,* Vol. 45, No. 5, pp. 205-214, May 1975.

Maul, L., "Noise immunity with Motorola high threshold logic," Motorola Integrated Circuit Application Note AN 298.

Mayer, F., and Faure, C., "Cables and Radio Noise Suppressors," *Rev. Gen. Elect.,* 75, pp. 307-313, March 1966 (ERA translation Trans. IB2402).

McDonald, G.M., "Instrumentation problems caused by common mode interference conversion," *IEEE Transactions on Electromagnetic Compatibility,* Vol. EMC-8, No. 1, pp. 17-24, March 1966.

Metcalfe, J.C., "The electromagnetic environment in cable tunnels," CEGB Marchwood Engineering Laboratories R/M/N 712, 1973.

Metcalfe, J.C., "Experimental investigation of electromagnetic induction in a cable installation," CEGB Marchwood Engineering Laboratories R/M/N/ 837, 1975.

Midwinter, J.E., "Trends in optical fibre transmission research," *Nature,* Vol. 261, June 3, 1976, pp. 371-376.

Miedzinski, J., and Pearce, S.F., "Interference from Industrial R.F. Equipment," ERA Report M/T 104, 1949.

Miedzinski, J., "Electromagnetic Screening, Theory and Practice," ERA Report M/T 135, 1959.

Ministry of Defence (MOD), U.K., "Code of practice (Ship Department)," TP 1006, 1972.

Mohr, R.J., "Induced noise in the reference ground of electronic systems," IEEE Electromagnetic Compatibility Symposium Record, 1970.

Monsanto, "Gallium Arsenide Optoisolators";
"GaAsLite Tips Vol. 2 (Principles of use)";
"Application information":
AN 510 Logic isolation line receiver
AN 500A Linear photodiode optoisolator circuits
AN 504 Using optoisolators in linear coupling
AN 505 Linear photodiode optoisolator circuits
MCD2-M Linear optoisolator data sheet for linear applications

Morrison, R., "Grounding and Shielding Techniques in Instrumentation," Wiley, 1968.

Mullard (a), "Transient voltage suppression with voltage regulator diodes," Application Note TP 1266, 1971.
(b), "Mullard to market solid state suppressors, six ranges up to 100 kW transient capability," Mullard Bulletin, September 1971.

Naito, H., and Shah, A., "Voltage induction by pulse-shaped Em fields on transmission lines, application to interference problems in electronic systems," CPEM Digest. Institute of Applied Physics, ETH, Zürich, 1974.

Rhodes, J.C., "The computer-instrument interface," *Control Engineering,* pp. 105-109, September 1966.

Rogers, A.J., "Fibre-optical communication," CERL Report No. RD/L/R 1908, 1975.

Sandbank, C.P., "The challenge of fibre optical communication systems," *Radio and Electronic Engineer,* Vol. 43, No. 11, November 1973, pp. 665-674.

Schelkunoff, S.A., "The impedance concept and its application to problems of reflection, refraction, shielding and power absorption," *BSTJ,* January 1938.

Schelkunoff, S.A., "Electromagnetic waves," D. Van Nostrand, New York, 1943.

Schlicke, H.N., and Struger, O.J., "Electrical interference in solid state industrial control systems," Allen Bradley Co., Milwaukee and Cleveland, U.S.A. (Upper Midwest Electrical Trade Show, Minneapolis, March 1972).

Seljeseth, H., "Grounding in E.H.V. stations." CIGRE Paper 36-77 (WG04) 05E, 1977.

Siemens, "Radio interference suppression—general," "Interference suppression units," Extracts from third volume of book series dealing with passive components and radio-interference suppression, Siemens A.G., Werk für bauelemente, vertrieb, 8000 Munich, 80, Balanstrasse 73, 1969/70.

Sladek, N.J., Electro-technology, Science and Engineering Series No. 95, Institution of Electrical and Electronics Technician Engineers, November 1960.

Tait, D.A.G., *Electronics Industry,* January 1978, pp. 17-21.

"Teleprotection," CIGRE 1969 (book), CIGRE, 112 Boulevard Haussmann, 75008, Paris.

Turner, H.J., "Field investigations of overvoltages on low-voltage wiring," CERL Report No. RD/L/R 1509, 1968.

Taylor, R.E., "Radio frequency interference handbook," NASA-SP-3067, 1971.

U.S.A.F., "Instrumentation grounding and noise minimisation handbook," Technical Report No. AFRPL-TR-65-1, 1965. Air Force Rocket Propulsion Laboratory, Research and

Technology Division, Air Force Systems Command, Edwards Air Force Base, California (from U.S. Department of Commerce, Defense Documentation Center, or British Library Boston Spa.) (fiche reference AD612027 RPL.TR.65.1, 5 fiches).

Vogl, W., "Problems connected with grounding and shielding in data processing systems." CIGRE Paper 36-77 (WG04) 04E, 1977.

Wheable, D., and Outram, S.H., "Series mode rejection in integratory digital voltmeters," IEE Conference Publication No. 65, pp. 70-73, 1970.

Wilson, I., "Electrical Interference," Technical File No. 23, *Engineering*. The Design Council, November 1975.

Wiszniewski, A., and Marczonek, S., "Transverse transient overvoltages and signal-to-noise ratio of secondary protection wiring," IEE Conference Publication No. 125, March 1975.

15 Bibliography

Where a serial number follows the reference, this applies to "Computer and Control Abstracts," and is shown in brackets thus: (8235).

For other Bibliographies see Greathead, Healey, Telephone Line and Teleprotection, which contain 231, 297, 82 and 67 references, respectively.

ACE Report 41 (1975) Electricity Council Guide to an ERA Code of Practice for the avoidance of interference with electronic instrumentation and systems 9PP. Internal use only. Indicates the thinking behind the guidance given in those aspects of the ERA Code of Practice relating to interference levels and susceptibility tests (Sections 3.4 and 9 of the Code of Practice) in respect of power-supply disturbances.

Adler and Kennedy, "Optical data links for industrial applications," *Instruments and Control Systems,* February 1972.

Aleksandrov, V.P., "Pulse noise immunity of a chain of logic elements," *Avtom. & Vychisl. Tekh.* (USSR), No. 5, pp. 65-70 (1971). In Russian. The concept Pulse Transmission Characteristic (PTC) is defined as the relationship between input and output pulse lengths. A method is proposed for determining the noise immunity of a chain of logic elements, based on the PTC. In worst-case calculations, noise immunity is determined graphically from the envelope of a family of PTCs. The Monte Carlo method is used for statistical modelling. A comparison is made of the graphical and statistical methods (8235).

American National Standard. "Guide for surge withstand capability (SWC) tests." ANSI C37 90a-1974, IEEE Std. 472-1974.

Anders, R., and Campling, A.C., "Investigations into interference in substation and power station auxiliary cabling," CIGRE Paper 36-09, 1976.

Arnold, A.H.M., "The theory of sheath losses in single-conductor lead-covered cables," *JIEE,* Vol. 67, pp. 69-96, 1928.

BS 5260. "Code of practice for radio interference suppression on marine installations," BSI, 101 Pentonville Road, London N1 9ND, 1975.

Ball, E.H., Occhini, E., and Luoni, G., "Sheath over voltages in high voltage cables due to special sheath bonding connections," IEEE Paper 31 TP 65-128, 1965.

Basu, R.N., "Measurement of small signals in a noisy environment." Proceedings of the conference on electrical interference in instrumentation, London, England, pp. 109-14, 9-11 June, 1970. The problem of using a sensitive system in an environment of ever increasing noise is defined. The main types of noise are named with examples. The protective measures against different kinds of interference are indicated with slight elucidation. The measurement of interference by a R.L. meter in this context is briefly mentioned. The types of shielding and their function are discussed in detail. A systematic approach to the design of shielded housing is delineated. Some guidelines based on practical experience are given for designing a lightweight low-noise instrumentation housing using modern technologies (495).

BEAMA, "Specification Guide, Electronic Equipment for Industrial Systems." Publication 209/2, May 1969, The British Electrical and Allied Manufacturers Association Ltd.

"Bibliography on radio frequency interference." *Trans. IRE,* RFI-4, No. 1, February 1962.

Birch, F.H., Burrows, G.H., and Turner, H.J., "Investigation into transient overvoltages on secondary wiring at switching stations," CIGRE Paper 31-04, Pt. II, pp. 9-15, 1968.

Bird, P. (ITT Semiconductors, Foots Cray, England). "A novel shaper circuit for d.t.l. and t.t.l. input interfacing," *New Electronics* (GB), Vol. 7, No. 13, pp. 28, 32 (25 June 1974). A versatile interfacing circuit is described for use when a t.t.l. or d.t.l. logic system must be driven from input signal sources that do not present the correct logic levels or are noisy. The circuit operates from +5 V, and is capable of being used as a level detector, noise suppressor, wave shaper or a.c. signal processor (no refs.).

Borgvall, T., Holmgren, B., Sunden, D., Widström, T., and Norback, K., "Voltages in substation control cables during switching operations." CIGRE Paper 36-05, 1970.

Bridges, J.E., "Study of low-frequency fields for coaxial and twisted pair cables," Proceedings of the 10th Tri-Service Conference on Electromagnetic Compatibility, Chicago, pp. 475-495, 1964.

British Electricity Boards, "Guide to an ERA Code of Practice for the avoidance of interference with electronic instrumentation and systems," ACE Report No. 41, the Electricity Council, 1975.

Brown, C.W., and Picken, D.A., "The use of fibre optic light guides in securing low interference measurements." Proceedings of the conference on electrical interference in instrumentation, London, England, 9-11 June 1970 (London, England, IEE 1970), pp. 37-39. The availability of fibre-optic light guides, light-emitting diodes and detectors has led to the development of a telemetry system for wide-band interference-free measurements in high-power circuits. Telemetry links applied to the simultaneous monitoring of a number of points in a 150 kV circuit enable interference-free oscillograms to be taken readily under transient conditions. The system also overcomes difficulties in making low-level measurements in the presence of high common mode voltages. Other uses for light guides in high-power systems include power-line current measurements and provision of control links with good electrical isolation (496).

Buisson, J., "Susceptibility of reactivity control instrumentation to electrical interference. Development of new cables." *CEA bulletin d'information scientifique et technique,* No. 195, pp. 13-17, September 1974. In French.

Bull, J.H., "Code of Practice for the avoidance of electrical interference in electronic instrumentation and systems," Electrical Research Association, Leatherhead, Report ERA 75-31, March 1975 (221 references).

C.C.I.T.T., "Directives concerning the protection of telecommunication lines against the harmful effects from electricity lines," I.T.U., Geneva, 1963.

Carter, G.W., "The electromagnetic field and its engineering aspects." Longmans Green, 1954.

CEGB, "Design Memorandum Draft Document 709/F/2, Cabling practice for voltages up to and including 11 kV," 1970.

CEGB, See "Inductive coordination."

CEGB, See "Telephone line protection."

Chowdhuri and Zobrist, "Susceptibility of electrical control systems to electromagnetic disturbances," *Transactions of the IEEE,* Vol. IA-9, No. 5, September/October 1973.

CIGRE, See "Teleprotection."

Cohen, A.A., "Long-distance digital interface system overcomes noise," *Electron. Engineering* (GB), Vol. 45, No. 542, p. 57-9, April 1973. Describes economical and reliable transmitter and receiver circuits, using i.c.s and pulse transformers, that are capable of interfacing digital systems over a distance of 1.5 miles in a noisy environment. The carrier signal of the modulated signal is 1 MHz, but the circuits will operate at higher frequencies if better-quality cables are used (3 refs.) (13729).

Coram, R.W., and Jackson, G.A., "Electrical interference in instrumentation: modern measuring techniques." Proceedings of the conference on electrical interference in instrumentation, London, England. IEE Conference Publication No. 65, 1970, pp. 18-22. Reviews techniques of measurement of the voltage and current generated by all sources of interference and also describes methods of determining the susceptibility of instruments and equipment to interference. Measuring equipment for the control of radio interference is relatively well known, and the application of this equipment to the general problems of interference with instrumentation is reviewed. Current design trends are concerned with the development of small, portable, self-contained measuring sets and with automatic test facilities. Details are given of measuring equipment currently being developed at ERA (497).

Cottet, Y., "Problems of measurements and the transmission of signals." International seminar on automatic control in lime, cement and connected industries, Brussels, Belgium, 9-13. Inst. Belge Regulation et Automatisme, 1968, pp. 293-298, September 1968. In French. The paper attempts to explain how, as we go further with automation, it has become necessary to improve the accuracy of any measurement and to reduce the error occurring during the remote transmission of control signals. Any control signal carries extraneous signals which disturb the original. A survey is made of all this noise or interference occurring on transmission lines. The various ways of reducing these disturbances are studied, such as use of grounded shielded wires, and oscillograph testing with introduction of filters. "Systems engineering in the glass industry" (8147, 7527).

Craven, A.T., "Electricity and magnetism for electrical engineers," Pitman, 1966.

Cruddace, A.E., and Starke, J.E., "Induction problems when using protection with correctly and incorrectly jointed underground pilot cores," IEE Conference Publication No. 125, pp. 152-159, 1975.

Damaye, R., "The sensitivity to parasitics of logic integrated circuits." *Electronique Industr.* (France), No. 125, pp. 479-83, July 1969. In French. The principal market for logic IC has been, from the start, that of computers; but today these circuits are extending their province to all branches of automation. It is mainly to satisfy these users that some IC manufacturers have undertaken, usually with success, the design and production of lines with high noise immunity. The meaning of this term is discussed (11186).

Del Noce, F., De Gasperi, G., Ghezzi, E. (Praxis, Milano, Italy). "Design of a logic circuit family for industrial applications with high dynamic noise rejection intrinsic safety and high reliability." Proceedings of the 11th automation and instrumentation conference (Milan, Italy: Fed. Sci. and Tech. Assocs. 1970), pp. 572-579, 23-25 November

1970. In English. The design of a logical circuit family for industrial applications stated to have the following basic characteristics is examined: high dynamic noise rejection, intrinsic safety, high reliability. High dynamic noise rejection: the insensibility to impulse disturbances introduced into the circuit, with the consequent possibility of use particularly suitable in industrial situations, is described. Intrinsic safety: the intrinsic safety of the circuits as an essential characteristic decided in the definition phase of the design is described. Emphasis is then given to the choice of components, to be made in conformity with the appropriate existing regulations. High reliability: the construction of the circuits with the use of highly professional components is described. In particular, the need is shown for subjecting the components to thorough quality testing and preaging. Finally, the description of assembly and final testing emphasises the guarantee of achieving high reliability in the final product.

Delorenzo, J.T., "Measurements of the shielding effectiveness of coaxial and two-conductor cables," *Review of Scientific Instruments,* Vol. 43, No. 1, pp. 161-164.

Dixon, J.F., "Microelectronics and digital techniques used in modern communications equipment," CEGB Advanced Telecommunications Conference, Buxton, March 1974, reprinted 1976, Transmission Division CEGB.

Dixon, J.F., "Report on microprocessors/microcomputers," ref. DT/JFD/573.10.430. April 1975, reprinted April 1976, Transmission Division CEGB.

Dvorak, T., "Electromagnetic field immunity—a new parameter in receiver design." *IEEE Transactions,* Vol. EMC-16, No. 3, August 1974.

Electrical Research Association, Cleeve Road, Leatherhead, U.K., "Interference with electronic instrumentation and systems," Reference 3041/68.

"Electromagnetic compatibility guide TP1006." Ministry of Defence. Director General Ships, 1 Pulteney Street, Bath, Somerset, BAI 5AB, 1972.

Fallou, M., "Analysis of methods of loss elimination in the metallic sheathing of single core a.c. cables," *Rev. Gen. Elect.,* Vol. 74, No. 6, pp. 481-489, CE Translation 4916, 1965.

Fillenberg, "Exploration of transients by switching capacitors," *Transactions of the IEEE,* Vol. PAS-90, No. 1, January/February 1971.

Fisher, F.A., *et al.,* "Transient Control Levels. A proposal for insulation coordination in low-voltage systems, *Transactions of the IEEE,* PAS-95, 1, pp. 120-130, January/February 1976.

Fleischhammer, W., Schneider, G., and Koppe, G. (Siemens AG, München, Germany). "T^3L: improved TTL circuits with incresed noise immunity," *Elektronik* (Germany), Vol. 23, No. 7, pp. 235-237, 246 (July 1974). In German. Describes a new TTL family, developed jointly by Siemens and RCA under the name T^3L, in which an increase in switching threshold leads to high noise immunity in both logic states, bringing it in line with CMOS. High-speed versions with Schottky diodes are also available. This new logic family is particularly useful in large data-processing systems (21952).

Fleischhammer, W., Schneider, G., and Koppe, G. (Siemens AG, Munich, Germany). "T^3L achieves C-MOS noise immunity while retaining TTL speed." *Electronics* (USA), Vol. 47, No. 5, pp. 94-97, 7 March 1974. The authors discuss the improvements in TTL circuits using a third transistor to shift threshold voltage to increase noise immunity. The circuit of a NAND gate is shown for both TTL and T^3L. The compatibility of the two logic families is shown, and applications of T³L are discussed. The threshold voltage and current/voltage characteristics are shown by graphs, and the family of ICs is shown (no refs) (11904).

Fowler, E.P., "On the interference immunity of coaxial cables." Eurocon 71, IEEE Region 8 Conference at Lausanne, October 1971.

Fowler, E.P., "A comparison of the screening performance of braided coaxial and triaxial cables, AEEW-R 982, March 1975.

Fowler, E.P., "Some recent work on the screening efficiency of braided and multi-screened signal cables." IEE Conference Publication No. 65 (Electrical Interference in Instrumentation), pp. 23-29, 1970.

Fowler, E.P., and Cox, A.E., "The practical application of electrical interference methods," IEE Conference Publication No. 65 (Electrical Interference in Instrumentation), pp. 60-69, 1970.

Fowler, E.P., and Gregory, H., "Studies of the source of electrical transients and interference at Wylfa Nuclear Power Station," IEE Conference Publication No. 65 (Electrical Interference in Instrumentation), pp. 81-89, 1970.

Fowler, E.P., "Interference immunity tests for nucleonic instrumentation." UKAEA Report AEEW R919, 1974.

Gillies and Ramberg, "Methods of reducing induced voltages in secondary circuits." *Transactions of the IEEE,* Vol. PAS-86, No. 7, July 1967.

Goldenberg, H., "The calculation of continuous current ratings and rating factors for transmission and distribution cables," ERA Report F/T187, 1958.

Gooding, F.H., and Slade, H.B., "Shielding of communication cables," *Trans. AIEE Communications and Electronics,* Vol. 74, pp. 378-387, 1955.

Grabe, H.F., "Noise level in sensitive measurement installations," *Elektro-Anzeiger* (Germany), Vol. 22, No. 23, p. 451-456, October 1969. In German. The article deals with the type and causes of fluctuation voltages, their influence on sensible measurement installations and the means to suppress them. The measurement installations considered include electronic amplifiers connected to multiple measuring devices, data-processing and control installations. The operation mode of the electronic amplifiers assumes a voltage-sensible input (24 refs.) (7245).

Greathead, A.P., "References on protection." R11.1.1 Memorandum 5/75, June 1975. P.O. Research Station Martlesham. (Contains 231 references subdivided to cover effects of lightning on cable systems, electronic-device failure and protection methods. Although aimed at telecommunication-system protection, many of the references are relevant to the subject of interference in electricity system stations.)

Gregory, H., "Interference measurement at Wylfa Power Station," CERL Report No. RD/L/N 15/70, 1970.

Grover, F.W., "Inductance calculations," Dover Publications, Inc., 1962.

Guillot, J., "Etude des perturbations electriques agissant sur les circuits BT des postes et centrales," *R.G.E.,* Tome 82, No. 11, November 1973.

Hall, B., *et al.,* "The effects of power-supply transients on British Post Office Cable Communication," *P.O.E.E.J.,* Vol. 67, Part 4, pp. 240-243, January 1975.

Halligan, J., "CMOS: a new logic type for control systems,"

Contr. Eng. (U.S.A.), Vol. 19, No. 6, pp. 68-70, June 1972. CMOS complementary metal oxide semiconductors circuits offer high power immunity, plus the additional benefits of operation over a broad range of power-supply levels and very low power dissipation. This article compares CMOS to other logic types and then describes how to interface it to them (22294).

Halman, T.R., and Harris, L.K., "Voltage surges in relay control circuits," *Transactions of the AIEE,* Vol. 67, 1948.

Hammond, P., "The magnetic screening effect of iron tubes," IEE Monograph No. 144, *JIEE,* Vol. 103, Pt. C, pp. 112-120, 1955.

Harrison, D., "A study of electrical interference in nucleonic reactor control instruments," UKAEA report AERE-RS/L14, 1963.

Harrison, D., "The mechanism of interference pick-up in cables and electronic equipment with special reference to nuclear power stations." *Radio and Electronic Engineer,* Vol. 29, No. 3, pp. 149-158, 1965.

Harvey, S.M., and Ponke, W.J., "E.M. shielding of a system computer in a 230 kV substation." IEEE F75 422-4, July 1975 (System Monitoring Division, Rockwell International, Anaheim, California).

Healey, Mrs. A.M., "Bibliography of electrical interference in instrument and control engineering." Corporate Development Laboratory, British Steel Corporation (covers 1965-72 and includes 297 selected references; see Chapter 14 references for details).

Hicks, R.L., "Control of transient interference on station wiring." Relay Committee, Pennsylvania Electric Association, Johnstown, Pennsylvania, USA, October 8, 1970.

Hicks, R.L., and Jones, D.E., "Transient voltages on power station wiring." IEEE Conference paper 70 TP 164-PWR, N.Y., January 1970.

Ikrath, K., "Leakage of electromagnetic energy from coaxial cable structures." Proceedings of the 4th Conference on Radio Interference Reduction, Armour Research Foundation, Chicago, October 1958.

"Inductive coordination between power and telecommunication lines." C.E. Bibliography No. 207, 1968 (contains 113 references) (obtainable from Reports Library, CEGB, Grindall House, 25 Newgate Street, London EC1).

"Interference suppression." BR 2120 (1957), MOD Director General Ships, 1 Pulteney Street, Bath, Somerset, BA1 5AB (see also Roessler, Amos, Goulter and Batt).

Jelinek, J., "An increase in resistivity of TTL logic circuits against noise." *Sdelovaci Tech.* (Czechoslovakia), Vol. 21, No. 20, pp. 375-376, October 1973. In Czech. Noise signals are divided into two groups according to their duration with respect to the speed required from logic circuits. From this point of view, the static and dynamic resistivities of TTL logic circuits are characterised. The dynamic resistivity is determined by the circuit delays. The functions of several flip-flop circuits are outlined, and their protection against noise signals by suitable selection of resistor and capacitor values is discussed. The protection of TTL circuits using the Miller integrator is mentioned (6 refs.) (9124).

Jesse, G. (Hartmann u. Braun AG, Frankfurt, Germany), "Localisation and elimination of electrical disturbances in measurement and control equipment." *Mess. & Pruef.* (Germany), No. 2, pp. 95-98, February 1974. In German. The mechanism, localisation and reduction of pick-up voltages due to stray inductive and capacitive coupling in equipment are explained and illustrated. Screening is discussed (10119).

Jones, F., "The shielding factor of underground cable sheaths," IEE Monograph No. 59, *Proceedings of the IEE,* Vol. 100, pt. IV, No. 1, pp. 34-38, 1953.

Kaden, H., "Wirbelströme und Schirmung in der Nachrichtentechnik" ("Eddy currents and screening in communications technology"), Springer-Verlag, Berlin, 1959.

King, L.V., "Electromagnetic shielding at radio frequencies." *Philosophical Magazine,* Vol. 15, Series 7, February 1933 (see also Kaden, H.).

Klipec, B.E., "Reducing electrical noise in instrument circuits," *IEEE Transactions,* Vol. IGA-3, No. 2, March/April 1967.

Kotheimer, W.C., "Shielding and grounding control cables. One technique for reducing surge voltages in control circuits." *Transmission and Distribution,* July 1974.

Laborelec (a), "High frequency interference in substations' auxiliary wiring." Laborelec (in English), 2-BJ/cd/204.76, 6 May 1974.

Laborelec (b), "Analyse de problèmes particuliers de perturbations." Laborelec 2.204 21-G2-BJ-mb, 8 June 1971.

Laborelec (c), "Le problème des perturbations dans les grands ensembles electroniques d'acquisition et de traitement de l'information." Laborelec 2.204-20-G2-RW, 15 May 1971 (CEGB Translation T4381).

Laborelec (d), "Etude des perturbations induites dans les circuits auxiliaires de centrales et de postes. Modes de generation—calcul—Moyens de reduction," Laborelec 2.204-19-G2-BJ, 8 June 1971.

Laborelec (e), "Perturbations induite dans les circuits

auxiliaires du poste 70/11 kV de Heppignies." Laborelec 2.204-47-BJ, 18 June 1973.

Laborelec (f), "Perturbations induites dans les circuits de transmission et de traitement de l'information." Laborelec 2.204-22-62-BJ, 2 June 1971.

Laborelec (g), "Perturbations induites dans les circuits auxiliaires du poste 70/11 kV de Heppignies. Resultats de mesures effectuees lors de defauts provoques." Laborelec 2-204-47-BJ, 18 June 1973, 40 (CEGB Translation T4417).

Lammeraner, J., and Stafl, M., "Eddy currents." Iliffe, 1966.

Lebedev, V.I., "Transfer functions of diode-transistor computer elements for calculating the limits of noise immunity." *Izvestia VUZ Radioelektronika* (USSR), Vol. 11, No. 12, pp. 1271-1278, December 1968. In Russian. The expression for the noise immunity, obtained in an analytical form, is examined as a function of circuit potentials, circuit resistances, number of loads, fixed and reverse transfer coefficients of the base current and the thermal currents. By reducing the thermal current by an order of magnitude, the noise immunity is improved by some 15 per cent. The temperature factor of the noise-immunity margin is established; the spread of the threshold and residual potentials, as well as the root-mean-square variation of the noise immunity, are calculated. A reduction of the bias potential, by about three times, degrades the noise immunity by 17 percent (6825).

Lyons, W., "Experiments on electromagnetic shielding at frequencies between one and thirty kilocycles," *Proceedings of the IRE,* Vol. 21, pp. 574-590, 1933.

Martin, D.J.R., "A general study of the leaky feeder principle," *The Radio and Electronic Engineer,* Vol. 45, No. 5, May 1975, pp. 205-214.

Martin, R.E., "The protection of telecommunication equipment from damage by overvoltages." CERL Laboratory Memorandum RD/L/M 187, 1968 (references to bibliographies of some 200 papers on inductive coordination and protection).

Master, C.A., "Electromagnetic interference control considerations, solenoid operated control devices." *IEE Trans. Electromagnetic Compatibility* (USA), Vol. EMC-11, No. 2, pp. 53-57, May 1969. Relay and solenoid controlled devices have historically been a common source of electromagnetic interference. The pulsed electrical energy resulting from actuation and deactuation generates broad frequency spectrum interference. Establishment of meaningful component and subsystem specifications and application of proper design techniques can minimise the interference generated by such devices and subsystems and can also minimise their susceptibility to other interference sources. This paper reviews the approaches which will significantly minimise electromagnetic interference problems (10308).

McCann, M., "Worst case TTL line driving," *Microelectronics* (GB), Vol. 3, No. 10, pp. 45-55, Spring 1971. Long signal interconnections between logic circuits produce pulse distortion due to the line mismatch. The resultant ringing reduces the noise margin to an extent depending upon the d.c. characteristics of the gate and the characteristic impedance of the interconnections. The noise margin is calculated for driving with standard and buffer TTL gates and receiving with a TTL gate with the addition of an input clamping diode (20067).

Mellitt, B., "Transient voltages generated by inductive switching in control circuits." (University of Birmingham, England). *Proc. Inst. Electr. Eng.* (GB), Vol. 121, No. 7, pp. 668-676, July 1974. In certain control applications, a common d.c. supply is used for both electronic and electromechanical equipment. In this situation, semiconductor devices may be subjected to transient voltages generated by the operation of mechanical switches in inductive circuits. This paper describes the factors involved in the generation of such transients, and methods for predicting their time duration are proposed. The circuit interruption process is described in terms of either stable or unstable arc switching, and the limiting factors for these two conditions are discussed. It is shown that, when unstable arc switching occurs, short-duration voltage spikes appear across the common power supply and, by implication, across all parallel-connected equipment. Reasonable correlation is shown to exist between predicted and measured transient-voltage time durations, and the analytical models proposed could be used as a theoretical basis for transient-voltage specifications (11 refs.) (19098).

Metcalfe, J.C., "The electromagnetic environment of cable tunnels," CEGB Report No. R/M/N 712, Marchwood Engineering Laboratories, 1973.

Metcalfe, J.C., "An assessment of the electromagnetic interference produced in a prototype cable containing both power and instrumentation circuits." CEGB Report No. R/M/N 706, Marchwood Engineering Laboratories, 1973.

Metcalfe, J.C., "Experimental investigation of electromagnetic induction in a cable installation." CEGB Report No. R/M/N 837. Marchwood Engineering Laboratories, 1975.

Miller, P., (SGS-ATES Agrate, Milan, Italy), "Is noise disturbing your logic?" *Electron* (GB), No. 28, pp. 89-91, 17 May 1973. The author discusses the use of logic families with high noise immunity and the simple ways of designing the circuits so that they are most effective. The two logic families are complementary m.o.s. (c.o.s./m.o.s.) and high-level logic (h.11) (no refs.) (20865).

Miller, P. (SGS—ATES, Milan, Italy), "Logic circuits operating under parasitical environmental noise conditions."

Toute Electron. (France), No. 383, pp. 30-33, December 1973. In French. Logic circuits, using semiconductors, are becoming more and more extensively employed industrially for such techniques as process control. In industrial environments, such equipments become subjected to considerable parasitic noise due to high-frequency generators, switching relays and the like. There are two apparent remedies for this troublesome interference. One is to isolate completely the affected circuits, the other to ensure that logic components having a high immunity to noise are used. Dependent upon the system of logic and the exact nature of the parasitic noise, either or both solutions may be employed. This article gives detailed consideration to two particular logic systems, and their immunity to high parasitic noise levels, and illustrates in a simple manner the aspects of circuits which give the most desirable efficiency (14292).

Mills, A.H., "Regional electromagnetic compatibility symposium record," San Antonio, Texas, USA, 6-8 October, IEEE, 1970, 5 pp. Complete characteristics of r.f. noise polluting the air waves cannot be obtained using standard off-the-shelf instrumentation. Generally, an amplitude-versus-frequency relationship does not adequately describe the noise. A statistical description better characterises this interface. An instrumentation system which provdes this description is discussed in this paper.

M.O.D. See "Interference suppression."

M.O.D. See "Electromagnetic compatibility guide," Roessler, Amos, Goulter and Batt.

Morrison, R., "Grounding and shielding techniques in instrumentation," Wiley, 1976.

Mosch, R., "Overvoltage protection of telephone switching equipment containing electronic components." *Electrical Communication,* Vol. 49, No. 1, 1974.

Moullin, E.B., "Principles of electromagnetism," Clarendon Press, Oxford, 1954.

Müller, F.E., "Interference elimination in switching regulated power supplies." *Electrical Communication,* Vol. 51, No. 1, pp. 28-33, 1976.

Newbery, P.G., "Fuses for power semiconductor protection." *Electrical Review,* Vol. 198, No. 6, pp. 27-29, 13 February 1976.

Oakley, R.J., "Surface transfer impedance measurements—a practical aid to communication cable shielding design," 18th International Wire and Cable Symposium, Atlantic City, New Jersey, USA, 1969.

Petrov, P., "Direct current noise immunity criteria of logical gates." *Izv. Inst. Tekh. Kilbern* (Bulgaria), Vol. 12, pp. 107-117, 1970. In Bulgarian. Basic questions connected with determining the capability of accurate operation of operational logical gates and logical structures in the presence of disturbing signals are considered. The noise signals and d.c. noise immunity are defined. A generalised representation of static characteristic of a logical structure is suggested. The parameters characterising the logical-gates immunity for some definite noises determined on these proposed criteria are used to estimate the particular systems of logical gates (17587).

Petrov, P., "Direct current noise immunity criteria of logic gates." *Izv. Inst. Tekh. Kibem* (Bulgaria), Vol. 7, pp. 107-117 (1970). In Bulgarian. The noise signals and d.c. noise immunity are defined. A generalised representation of static characteristic of a logic structure is suggested. The parameters characterising the logic gates immunity for some definite noises are determined on this basis. Estimation criteria of d.c. noise immunity are stated. The relation between the d.c. noise immunity and the fanout of the logic gate is elucidated and suggested as a generalised estimation criterion of the logic-gate quality. The proposed criteria are used to estimate the particular systems of logic gates (24390).

Phelps, J.W., "Electrical protection for transistorised equipment." *Bell Laboratories Record,* July 1958.

Phillips, P. *et al.,* "Suppression of relay interference, with particular reference to nuclear reactor safety circuits," UKAEA Reactor Group AEEW-R 1031, 1975.

Rajda, F. (Power Res. Inst. for Energy, Prague, Czechoslovakia), "Interference voltage problems in power plant controlling computer circuits." Proceedings of the conference on electrical interference in instrumentation, London, England. IEE Conference Publication No. 65, 1970. The common mode rejection of differential amplifiers is solved theoretically and practically. The noise measurement system in power plants is analysed, and values measured on 32 thermocouple lines of the 200 MW set in Ledvice power plant are given. The measurement analysis contains theoretical and practical conclusions, and some problems of series mode filters are added. Further, the report deals with the systematic treatment of the generation of electrical interference voltage in measuring leads between sensors and the control computer of a power plant. an original theory has been worked out for computing the voltage due to capacitive and magnetic couplings for shielded and unshielded telecommunication cables.

Ramo, Whinnery and Van Duzer, "Fields and waves in communication electronics," Wiley, 1965.

Roessler, Amos, Goulter and Batt, "Radio interference suppression design guide." ASWE Laboratory Note Q-65-2, 1965 (MOD, Director General Ships, 1 Pulteney Street, Bath, Somerset, BAI 5AB).

Ruppenthal, P., "Noise immunity of logic circuits." *Hidadas-*

technika (Hungary), Vol. 24, No. 7, pp. 207-214, July 1973. In Hungarian. Electronic data-processing equipments are functioning more and more to a greater extent in an industrial environment of high interference level. Therefore, the noise immunity of the logic circuits becomes an important problem. After discussion of noise immunity in basic terms, the causes of interference and the possibilities of protection against interference are summarised. After the problems of noise immunity of TTL integrated circuits, integrated logic circuits of high noise immunity are dealt with. Methods by which the noise immunity of the integrated circuits can be increased by external circuit elements are illustrated (11 refs.) (3314).

Schaffner, Hans, "Line interference simulators NSG200," "Impulse voltage generator NSG504," "Broadband RFI suppression filters for IC circuits." 1975 (three catalogues from Hans Schaffner AG, 4708, Luterbach, Switzerland).

Scheide, A.W., "Evaluation of the shielding properties of cable trays for use in an industrial environment," *IEEE Transactions on Industry Applications,* Vol. IA-8, No. 6, pp. 783-787, 1972.

Schelkunoff, S.A., "The electromagnetic theory of coaxial lines and cylindrical shields," *BSTJ,* Vol. 13, pp. 532-579, October 1934.

Schelkunoff, S.A., and Odarenko, T.M., "Crosstalk between coaxial transmission lines," *BSTJ,* Vol. XVI, April 1937.

Schlicke, H.M., and Struger, O.J. (Allen-Bradley Co., Milwaukee, Wisconsin, USA), "Getting noise immunity in industrial controls." *IEEE Spectrum* (USA), Vol. 10, No. 6, pp. 30-35, June 1973. Guidelines for noise-hardened design aid systems engineers in minimising equipment downtime (no refs.) (18461).

Scott-Kerr, R.J., "Computer-aided control at Pembroke Power

Station," *Electronics and Power,* Vol. 18, pp. 87-90, 1972.

Seljeseth, H., "Electrical noise in control equipment used for power supply and by industry—Norwegian E.F.I. Committee has draft code of practice for electronic equipment requirements. Part II." (In Norwegian.) *Elektro,* Vol. 89 (4), pp. 14-15, 26 February 1976.

Seljeseth, H., "Guidelines for design of control systems to minimise electrical noise from external sources in power plants and industry." The Norwegian Research Institute of Electricity Supply, E.F.I. TR No. 1774 E (in English). March 1974, revised edition TR 1774 B (in Norwegian). September 1975.

Simons, K.A., "Attempts to relate transfer impedance to radiation under somewhat realistic conditions." IEC Working Group Paper SC46A/WG1 (Simons), 3 September 1974.

Sonneman, W.K., "Transient voltages in relay control circuits." *AIEE Transactions,* Vol. 80, pt. III, pp. 1155-1162, 1962.

Stein, P.K. (Arizona State University, Tucson, USA), "The response of transducers to their environment—the problem of signal and noise." *Journal of Environmental Science* (USA), Vol. 14, No. 3, pp. 10-17, May-June 1971. Discusses the problem of how transducers respond to their environment so that only the desired environmental aspect is achieved. Since every link in the measuring chain is capable of producing three entirely different noise levels, the problem of their specific identification at any instant in any test situation becomes important. The emphasis here is on the documentation of these noise levels. The methods presented permit documented presence of noise to identify specific problems. Documented absence of noise guarantees the data, regardless of how inexplicable these data appear.

Stratton, J.A., "Electromagnetic theory." McGraw-Hill.

Sutton, H.J., "Transients induced in control cables located in EHV substations." *IEEE Transactions,* Vol. PAS-89, No. 6, pp. 1069-1081, July/August 1970.

"Telephone line protection against overvoltages." C.E. Bibliography No. 208, 1968 (contains 82 references) (obtainable from Reports Library, CEGB, Grindall House, 25 Newgate Street, London EC1).

"Teleprotection." CIGRE 1969 (book). Section 9.2.4, p. 271. Contains 67 references on induction, transients interference and communication line protection.

Toland, H., "The effect of radio frequency interference on electronic equipment at Hunterston B Power Station." South of Scotland Report No. BRT/75/39, June 1975.

Tomica, A., "Bridge method simplifies screening-factor measurement." *Electronic Engineering,* pp. 76-78, September 1970.

Towle, L.C., and Burkitt, C.J., "Some effects of low frequency interference when using thermocouples for industrial temperature measurement." Conference on electrical interference in instrumentation, London, England, 1970. IEE Conference Publication No. 65.

U.S.A.F., "Instrumentation grounding and noise minimisation handbook." Technical Report No. AFRPL-TR-65-1. U.S. Air Force Rocket Propulsion Laboratory. Research and Technology Division. Air Force Systems Command, Edwards Air Force Base, California (from U.S. Department of Commerce, Defense Documentation Center or British Library Boston Spa [Fiche Reference AD612027 RPL. TR-65-1, 5 fiches]).

Vasil'eva, N.P., Sedykh, O.A., and Fioshkina, O.M., "Providing maximum noise immunity of a system of logical elements with follower and inverter characteristics." *Avtom. & Telemekh* (USSR), Vol. 33, No. 2, pp. 150-161, February 1972. In Russian, English translation in: *Autom. and Remote Contr.* (USA), Vol. 33, No. 2, Pt. 2, pp. 296-305, February 1972. A method is considered for the choice of levels of the 0 and 1 signals, providing maximal noise immunity in systems consisting of logical elements with arbitrary and stiffly coupled F (follower) and I (inverter) characteristics. An example is given of the determination of optimal parameters of a system of ferrite-diode logical elements (2 refs.) (24417).

Vogel, O., and Hubensteiner, H., "Measures for limiting transient overvoltages in the measuring, control and signal wires of extensive hv outdoor stations." CIGRE Paper 34-01, 1974.

Wedmore, E.B., Morgan, P.D., and Whitehead, S., "A critical study of a three-phase system of unarmoured single-conductor cables, from the standpoint of the power losses, line constants and interference with communication circuits." *JIEE,* Vol. 67, 359-434, 1929.

Wavre, A. (Westinghouse Electric Corp., Research Lab., Pittsburgh, Pennsylvania, USA), "Application of integrated circuits to industrial control systems with high noise environments." Conference record of 1968 third annual meeting of the IEEE industry and general applications section, Chicago, Illinois, 29 September-30 October 1968 (New York: IEEE, 1969), pp. 255-263. A definition of both d.c. and a.c. noise margin is presented and used to compare the noise sensitivity of various logic devices. For integrated-circuit applications in very-high-noise environments, a systematic design method is proposed that produces noise-immune systems. The method, which is based on the flow table approach for sequential circuits, gives the location of a minimum number of slow gates or slow-down networks, while critical rate conditions are easily identified and eliminated (5589).

Willard, F.G., "Prevention and treatment of noise in control signals." *IEEE Trans. Industr. Gen. Applc.* (USA), Vol. IGA-5, No. 3, pp. 266-272 (May 1969). Electrical control signals transmitted between two points usually arrive more or less distorted by noise. If proper control action is to occur, the information contained in the signals must be recovered and the noise must be rejected. This recovery and rejection are accomplished by invoking the principle of contrast, discriminating between useful information and noise as a result of some characteristic difference between them. Various practical means to effect and employ such contrast are examined, and a number of selected examples are presented. The role of proper wiring layout and a grounding is stressed. High-level noise can do more than obscure information in a signal. A number of instances exist in which actual equipment damage has been caused by noise. A brief description, with specific illustrations, is presented (436).

Wilson, I., "Susceptibility testing on instrumentation installations." Eurocon 71, IEEE Region 8 Conference at Lausanne, October 1971.

Wilson, I., "Electrical interference, Technical File No. 23." *Engineering,* November 1975.

Wilson, I., and Fowler, E.P., "The design and use of high performance screened cables for reducing electrical interference effects in neutron flux measuring channels." (IAEA specialist meeting, Prague 1973). IAEA-SM-168/D-8, 1973.

Wilson, I., *et al.,* "Electrical interference in instrumentation systems." UKAEA, Winfrith, to be published.

Wiszniewski, A., and Marczonek, S., "Transverse transient overvoltages and signal-to-noise ratio on secondary protection wiring." IEE Conference Publication No. 125, pp. 304-309, 1975.

Yarushek, V.Ye., "Synthesis of circuits for processing digital signals." *Tekh. Kibernetika* (USSR). English translation in: *Engng. Cybernetics* (USA), No. 2, pp. 91-96, March 1969. A method is proposed for the abstract synthesis of logic circuits intended for the recognition of signals with a noise background using statistically independent readings of a receiver output signal, quantised to an arbitrary number of levels. The method permits reduced disjunctive forms of the logical functions describing the statistically optimal decision device to be obtained from known conditional distributions of the quantised signals (20590).

Zayezdnyi, A.M., and Fink, L.M., "The problem of electromagnetic compatibility." *Telecommunications and Radio Engineering,* Pt. 2 (USA), Vol. 28, No. 7, pp. 59-63, July 1973. Translation of: *Radioteknika, Moskva* (USSR), Vol. 28, No. 7, pp. 2-9, July 1973 (received July 1974). The problem of electromagnetic compatibility is shown to be directly linked with aspects of multilayer multiplexing and the nonlinear separation of signals. It is shown that the specific carrying capacity and noise immunity can be increased by using new methods of signal multiplexing and processing (21 refs.) (18845).

APPENDIX A

Addresses of Manufacturers Referred to in the Text

Inclusion of a manufacturer's name does not imply that the products concerned are approved, nor that other manufacturers not listed may not offer alternatives.

1. Alma Components Ltd., Park Road, Diss, Norfolk (0379 2289).

2. Altoflex, 36 Rue de Maraichers, F 75020, Paris (628-46-85).

3. American Laser Systems, Inc., 106 James Fowler Road, Santa Barbara Airport, Goleta, California 93017, USA (805-967-0423).

4. Amphenol Ltd., Warwick House, Azalea Drive, Swanley, Kent BR8 8HL ([0322] 68424, Telex 896520).

5. Analog Devices, Inc., P.O. Box 280, Route 1, Norwood, Massachusetts 02062, USA, and Central Avenue, East Molesey, Surrey, England.

6. Asea-hafo, Fack 162-10 Vallingby, Sweden (see also MCP Electronics).

7. B&R Relays Ltd., South Road, Temple Fields, Harlow, Essex (0279 34561).

8. Beckman Instruments, Queensway, Glenrothes, Fife, Scotland, KY7 5PU (0592 753811).

9. Belling & Lee Ltd., 540 Great Cambridge Road, Enfield, Middlesex, EN1 3RY (01 363 5393).

10. Borer Electronics Ltd., Longfield Road, Twyford, Berks, RG10 9AU (0734 340416, 0734 341414).

11. Brown Boveri Co., Glen House, Stag Place, London SW1E 5AH (01 828 9422).

12. Burr Brown, P.O. Box 11400, Tucson, Arizona 85734 ([602] 294-1431) ((Telex 910 952 1111).

13. Cerberus Ltd., Mannedorf, CH 8708, Switzerland.

14. Chomerics Europe, 66 Rue de la Boetie, 75008 Paris, France.

15. Clare International, Rue General Gratry 102, 1040 Bruxelles, and Malt House, Field End Road, Ruislip, Middlesex, HA4 9PL (01 868 3533).

16. Control Technology Ltd., Bolney Avenue, Peacehaven, Sussex, 3N9, 8HQ (079 14 584).

17. Cooknell Electronics Ltd., Weymouth, Dorset (Weymouth 73744).

18. Coutant Electronics Ltd., Trafford Road, Reading, RG1 8JR (0734 55391).

19. Creative Instrumentation Ltd., Unit 4, Decoy Road, Worthing, Sussex.

20. Danbridge (UK) Ltd., Sherwood House, High Street, Crowthorne RG11 7AT, Berks. (034 46 2369).

21. Delle CGE, 95 Rue Frederic Fays, 69611 Villeurbanne, France.

22. Dubilier Electronics Ltd., Ashcroft Road, Kirby Industrial Estate, Liverpool, L33 7TR (05 546 3535) (Telex 62153).

23. Emerson & Cuming Europe, Nijverheidsstraat 24B-2431 Oevel, Belgium (014-589326) (Telex 31500).

24. Erie Electronics Ltd., South Denes, Great Yarmouth, Norfolk (0493 56122).

25. Gardners Transformers Ltd., Christchurch, Dorset, BH23 3PN (02 015 2284).

26. G.E.C. Rectifiers Ltd., Stafford ST17 4LN (0785-51222, Ext. 410) (Telex 36206).

27. General Semiconductor Industries, Inc., 2001 West Tenth Place, Tempe, Arizona 85281 (602-968-3101) (Telex 910.950.1942).

28. Hewlett Packard, Kingstreet Lane, Winnersh, Wokingham, Berks. RG11 5AR (Wokingham 784 774) (Slough 33341).

29. Hakuto Co. Ltd., CPO Box 25, Tokyo 100-91, Japan.

30. Harris Corporation, P.O. Box 37, Melbourne, Florida 32901, USA (305-727-4000).

31. Hird Brown Electronics Co., Lever Street, Bolton, Lancs. (Bolton 27311).

32. I.T.T. Component Group Europe, Standard Telephones & Cables Ltd., Optical Equipment Division, Edinburgh Way, Harlow, Essex CM20 2DE.

33. Intronics, 57 Chapel Street, Newton, Massachusetts 02158,

USA (617-332-7350, Telex 710-335-6835) (Hitchin).

34. K&N Electronics Ltd., Cordwallis Street, Maidenhead, Berks. (0628-22447) (Telex 84692).

35. Klippon Electricals Ltd., Power Station Road, Sheerness, Kent (Sheerness 3093).

36. Knitmesh Ltd., 36 Victoria Road, London W1.

37. Lambda Electronics, Abbey Barn Road, High Wycombe, Bucks. (High Wycombe 36386).

38. Friedrich Lutze, Postfach 1224, D7056, Weinstadt, Germany.

39. Lyons Instruments Ltd., Hoddeson, Herts. (Hod. 67161).

40. Manufacturing Processes Ltd., Belvedere Works, Bilton Way, Pump Lane Industrial Estate, Hayes, Middlesex (01 848 9871).

41. MCP Electronics Ltd., Station Wharf, Alperton, Wembley, Middlesex HA0 4PE (01 902 5941).

42. Measurement Technology Ltd., 26-30 John Street, Luton, Beds. (0582 33945/6).

43. M-O Valve Co., Shepherds Bush Road, Hammersmith, London W6 7PE (01 603 3431).

44. Meret, Inc., 1815 24th Street, Santa Monica, California 90404, USA (213-828-7496).

45. Monsanto Ltd., 10 Victoria Street, London SW1H 0NG (01 222 5678).

46. Mullard Ltd., Industrial Electronics Division, Torrington

Place, London WC1E 7HD (01 580 6633).

47. National Semiconductor Ltd., Stamford House, New Road, Altrincham, Cheshire, WA141 DR, England.

48. Nolton Communications Ltd., Delamore Road, Chestnut, Herts.

49. Optronic Fort Ltd., 32 Duncan Terrace, London N1 8BS (01 278 6825).

50. Pearson Electronics, Inc., 4007 Transport Street, Palo Alto, California (415-326-7285).

51. Procond S.p.a., 32013 Longarone (Belluno), Italy.

52. Rifa A.B., S161, 11 Bromma, Sweden (Stockholm [8] 26 26 00, Telex 10308) (Jermyn Distribution, Sevenoaks, Kent (0732 50144) (Telex 95142).

53. Roband Electronics Ltd., Charlwood, Surrey RH6 0BJ, England.

54. Rochester Instrument Systems, Maxim Road, Dartford, Kent DA1 4BG (032 25 26211).

55. Roditi International, Carrington House, 130 Regent Street, London W1R 6BR (01-437 1997) (Telex 265661).

56. Rosemount Engineering Co. Ltd., Durban Road, Bognor Regis, Sussex (02433 3121).

57. Roxburgh Electronics Ltd., 22 Winchelsea Road, Rye, Sussex (Rye 3777, Telex 957066).

58. Siemens Ltd., Gt. West Road, Brentford, Middlesex (01-568 9133) (Telex 23176).

59. Sprecher & Schuh U.K. Ltd., Thame, Oxfordshire (084421 3271).

60. Standard Telephones & Cables Ltd., 136 Corporation Road, Newport, Mon. NPT OTF (0633 56281).

61. The Steatite Group, Hagley House, Hagley Road, Birmingham B16 8QW (021 454 6961).

62. Stolec Ltd., 22 Winchelsea Road, Rye, Sussex (Rye 3777) Cosmocord-Stolec 0992-27331).

63. Syscom A.G., Rutlistrasse 50, CH-4051 Basel, Switzerland (061-227171).

64. Systems Engineering Services Ltd., 35-39 South Ealing Road, London W5.

65. Technical Wire Products, Inc., Stephenson Mills Ltd., St. John's Road, New Malden, Surrey.

66. Teknis Electronics, Shirley Lodge, 470 London Road, Slough, Berks. SL3 8QY (0753-44712) (Telex 847185).

67. Teledyne Ltd., Heathrow House, Bath Road, Cranford, Middlesex TW5 9QP (01 897 2501).

68. Tekdata Ltd., Westport Lake, Canal Lane, Tunstall, Stoke-on-Trent, Staffs. ST6 4PA (0782 811 711).

69. Toko U.K. Ltd., Ward Royal Parade, Alma Road, Windsor, England (07535 54057) (Telex 848095).

70. Unitrode Corp., 580 Pleasant Street, Watertown, Massachusetts 02172, USA.

71. Walmore Electronics Ltd., 11 Betterton Street, Drury Lane, London WC2M 9BS (01 835 1228).

72. Waycom Ltd., Wokingham Road, Bracknell, Berks. RG12 1ND (Bracknell 24571) (Schaffner filters, etc.).

73. Whiteley Electrical Radio Co. Ltd., Mansfield, Notts. NG18 5RW (0623 24762).

APPENDIX B

Electromagnetic Radiation and Its Interaction with the Screening of Cables

By E.P. Fowler, UKAEA, Winfrith

NOTE: Numbered references quoted in Appendix B are listed at the end of the Appendix itself.

B.1 Introduction

This Appendix outlines the parameters used to define cable screening and shows how they may interact with electromagnetic fields. No quantitative study is attempted, but mention is made of an investigation into the propagation of disturbances in an industrial environment and other studies on more open cables by other authors. It seems that we have some qualitative understanding of the problem but that there is much work yet to be done on the interaction of electromagnetic radiation and the shielding performance of coaxial cables expressed as it is in terms of its surface transfer impedance. Some problem areas are:

(a) The need to establish the maximum field radiated from a CATV cable both to meet regulations on radiated interference and to prevent unauthorised reception.
(b) The interaction between powerful radio and radar transmitters and nearby instrumentation cables. This problem is particularly acute in marine installations.
(c) A problem of growing importance is the disturbance of in-

dustrial instruments and cables by portable radio transmitters.

(d) Applications involving the deliberate radiation from cables as in leaky feeder systems developed for use on railways and in mines.

B.2 Parameters Used to Describe Cable Screening

The two parameters commonly used to express coaxial cable screening performance are Z_T (surface transfer impedance) and Z_F (capacitive coupling impedance). In many situations, a parameter complementary to Z_F must be considered which describes the direct influence of local magnetic field on a cable. In line with the preference for Z_F over Y_T (transfer admittance) or C_{12} (braid through capacitance), a symbol Z_{SH} (shine impedance) has been coined to describe the effect on a cable of current flowing in a nearby conductor but not returning through the cable sheath.

The three parameters Z_T, Z_F and Z_{SH} are all expressed in the same units (generally mΩ/m) and are described here as in a previous report (1).

B.2.1 Surface Transfer Impedance

Surface transfer impedance per unit length of a cable (Z_T) relates the voltage generated inside the sheath to the current flowing on the outside. If the cable sample is electrically short ($a<\frac{\lambda}{8}$, Fig. B.1), the voltage inside the sheath (V_1) induces a current in the circuit, and is therefore divided between the terminating impedances Z_1 and Z_2. The value of Z_T is normally quoted in mΩ/m, with $f=30$ MHz used as the reference frequency.

At low frequencies, Z_T is resistive and has a value equal to the sheath (or braid) resistance. At frequencies above 1 MHz, Z_T may vary considerably either way from its d.c. value owing to detailed construction of the braid, coupled with skin effect (which works to improve the screening).

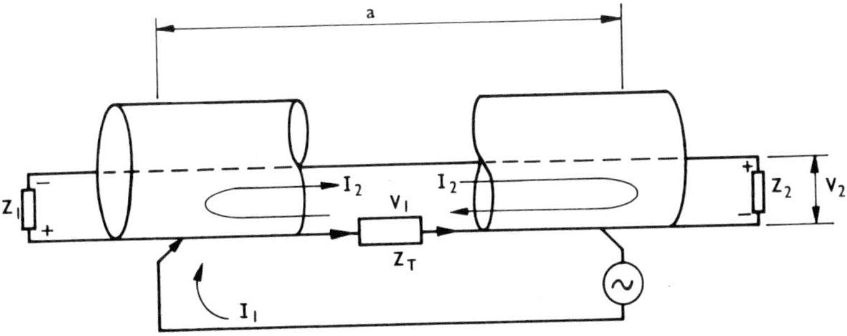

$$Z_T = \frac{2V_2}{aI_1}$$

Fig. B. 1
SURFACE TRANSFER IMPEDANCE Z_T

B.2.2 Capacitive Coupling Impedance

Capacitive coupling impedance per unit length (Z_F) effectively relates the current injected through the sheath to the electric field outside. The mechanism involved is direct capacitive coupling from conductors outside the screen to the inner conductor. The symbol for this through coupling capacitance is C_{12}; but it is often more convenient to use capacitive coupling impedance Z_F, as it is comparable with transfer impedance Z_T. Z_F is normally expressed in mΩ/m in terms of the current flowing in a correctly terminated injection line (Fig. B.2) to a correctly terminated cable sample. If the through capacitance (C_{12}) is measured, there is a simple expression for deriving Z_F:

$$Z_F = 2\pi f\, Z_{01}\, Z_{02}\, C_{12} \qquad \ldots(1)$$

where Z_{01} Z_{02} are the characteristic impedances of the outer and inner transmission lines.

B.2.3 Cable Eccentricity

The dimensional eccentricity of the inner conductor is difficult to measure in a coaxial cable, but its effect on cable screening can be expressed in terms of an equivalent transfer impedance to a magnetic field. A magnetic field (H) has units

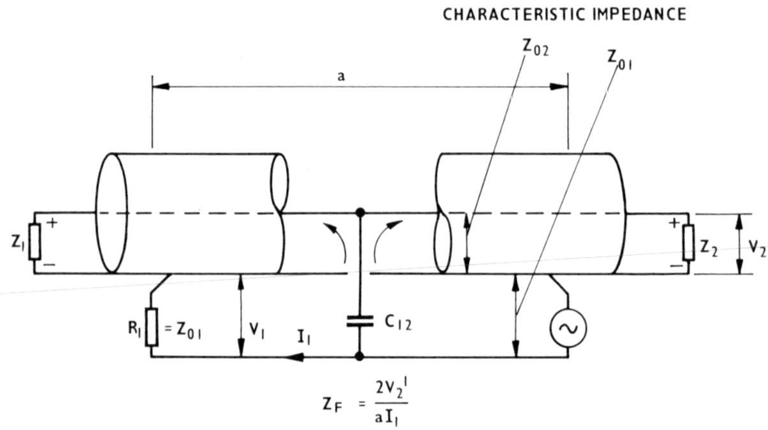

Fig. B. 2
CAPACITIVE COUPLING IMPEDANCE Z_F

A/m, so that equivalent transfer impedance to a magnetic field has units Ωm per unit length or simply Ω. Measurement is conveniently made with a current flowing in a nearby conductor (Fig. B.3). Shine impedance per unit length (Z_{SH}) is defined as the net voltage generated per unit length (or the voltage difference between inner conductor and sheath) for unit current flowing in a conductor spaced away by $r_o = 10$ mm. The term "shine impedance" was coined because attempts to measure the transfer impedance of screened enclosures were affected by the path the disturbing current was constrained to take in the screen. As there was greater coupling near holes, it was as though the current shone through gaps in the screen.

In Fig. B.3(A) Z_{SH} in Ω/m is given by:

$$Z_{SH} = \frac{V_1}{I_1 a} = \frac{2V_2}{I_1 a}$$

Although a coaxial-cable inner conductor has a low but indefinite eccentricity, an individual conductor in a multicore cable may have a large and definable eccentricity. Therefore we need to find the relationship between conductor eccentricity (D) and the value of Z_{SH}, and it may be useful to express the screening in terms of D. The symbols used are shown in Fig. B.3(B).

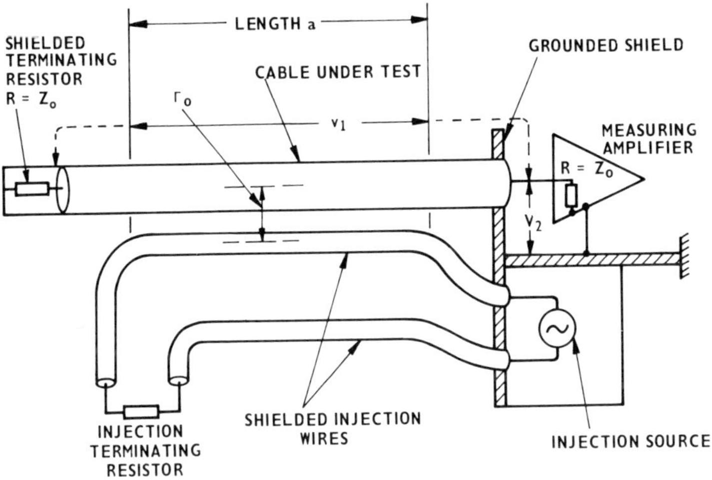

(A) SCHEMATIC OF TEST SET-UP

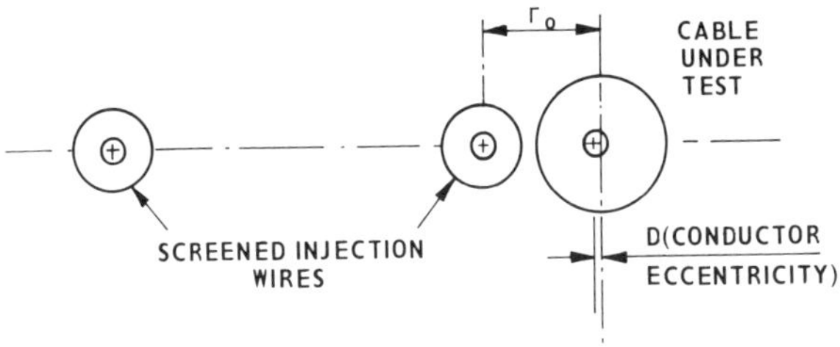

(B) SECTION OF CABLE AND INJECTION WIRE

Fig. B. 3

MEASUREMENT OF Z_{SH} FOR CABLE ECCENTRICITY

For a single wire carrying a current I, the magnetic field strength at a distance r is given by:

$$H = \frac{I}{2\pi r}$$

$$B = \mu_o H = \frac{\mu_o I}{2\pi r}$$

Assuming D is small compared with r, the total flux linking the circuit formed by the wire and screen over length a is given by $\Phi = B \times \text{area} = BDa$

$$\therefore \Phi = \frac{\mu_o \mathrm{IDa}}{2\pi \mathrm{r}}$$

Let $\mathrm{I} = \hat{\mathrm{I}} \sin \omega t$; the voltage V induced in the cable is given by:

$$\mathrm{V} = \frac{d\Phi}{dt} = \frac{\mu_o \hat{\mathrm{I}}\mathrm{Da}\ \omega \cos \omega t}{2\pi \mathrm{r}}$$

But $\omega = 2\pi f$; let $a = 1$ meter

$$\therefore \mathrm{V} = \frac{\mu_o \mathrm{IDf}}{\mathrm{r}} \text{volts/m} \qquad \ldots(2)$$

The shine impedance Z_{SH} is defined as the voltage generated for a current flowing in a nearby wire spaced by $r_o = 10$ mm. With the eccentricity D in meters:

$$Z_{SH} = \frac{\mathrm{V}}{\mathrm{I}} = \frac{\mu_o \mathrm{Df}}{r_o} = 100\ \mu_o \mathrm{Df} \qquad \ldots(3)$$

As an example, the possible value for Z_{SH} can be estimated from the tolerance on cable dimensions. A typical cable has diameter over the dielectric of 2.95 mm $\pm$0.12 mm. If the inner conductor eccentricity equals the tolerance of 0.12 mm, the value of Z_{SH} is found to be 1.5 mΩ/m at a frequency of 100 kHz.

B.3 Electromagnetic Fields

Associated with any alternating-current circuit are electric and magnetic fields surrounding the conductors. It is convenient to classify these as near fields and far fields to compare their properties. The near magnetic or induction field is generated by current flow and collapses back into the conductor each halfcycle. Similarly, the near electric field, generated by voltage difference, collapses each halfcycle. The energy in the far field does not collapse but is converted alternately from electric to magnetic and can be described therefore as a radiated field. The relative magnitudes of the two forms are controlled by the impedance of free space (=377 ohms).

Near field intensity is proportional to the inverse cube of the spacing from the generating source, while far field intensity is proportional directly to the inverse spacing. It is to be expected that the larger near field will give way to the far field at some distance. This generally happens at a spacing of about $\lambda/2$. For cables supported on poles, fences and the exterior walls of buildings (CATV applications particularly), the far field is likely to dominate for frequencies greater than 50 MHz and spacing greater than 3 m. In all other situations where there are nearby cables and other conducting metalwork, we are concerned with the near field and so may have to consider electric and magnetic effects separately.

B.3.1 Far Field Coupling

A coaxial cable illuminated by an electromagnetic field from a remote source will have voltages and currents induced in its screen, and their magnitudes are presumably calculable. The current flowing down the cable screen can be ascertained, and hence the disturbing voltage and its distribution along the cable can be derived in principle from its surface transfer impedance (Z_T). It can be assumed here that at those frequencies where we can possibly restrict ourselves to far field coupling the contribution from Z_F and Z_{SH} (shine impedance) can be ignored compared with that from Z_T. The calculations become more complex as the field becomes nonplanar and when the immediate neighbourhood of the cable is not a uniform dielectric. Such situations are the only ones of real interest, and it is here that some approximate relationship is required between free-space field strength and cable screening parameters.

A preliminary investigation by Simons (2) introduces the term "system shielding" to describe the relationship between a cable's surface transfer impedance and the field measured at a significant distance from it. Using a long cable and measuring at a predetermined spacing with a dipole, the system shielding records the increased coupling to the measuring dipole when the cable under test is replaced by another tuned dipole. He gives figures of 50 dB to 60 dB at frequencies between 50 MHz and 275 MHz for the system shielding of a cable with Z_T normalised to 1 Ω/m. The measuring dipole was spaced away by 3 m. These figures are especially useful for comparison with the near field situation.

It is interesting to note that over the restricted frequency band of his measurements Simons finds that the system shielding is not strongly dependent on frequency. As the measurement uses a tuned dipole of constant sensitivity to field strength, the radiation from a long cable must also be independent of frequency.

B.3.2 Near Field Coupling from a Radio Transmitter

Preliminary experiments have been set up to measure the disturbance from a hand-held portable transmitter in close proximity to a cable lying on a tray. Spacings of 30 cm and 3 cm were chosen using a transmitter radiating in the UHF band (~462 MHz). The disturbance induced from the transmitter antenna was compared with that from a 200 ohm terminated transmission line, known colloquially as a "tickle stick," operating at the same frequency when in the same position.

The terminated line was more effective as a disturbing source than the UHF aerial by a factor of 45 dB at 30 cm spacing and 28 dB at 3 cm spacing. The power supplied to the terminated line was 10 mW, while the designated power of the UHF transmitter was 1 W. It is not known whether the UHF transmitter was oriented for maximum coupling nor whether the antenna power was affected by its close proximity to a metal tray. However, it seems that much greater coupling was obtained with the tickle stick and that the use of this form of disturbing source is attractive for immunity testing if only because of its greater efficiency in near field conditions and its aperiodic nature.

The leaky feeder principle, when used for communication in tunnels and other enclosed spaces, is an application of near field coupling. Studies of coupling loss between a cable and a nearby antenna while both are within a tunnel show it to be markedly dependent on the ratio of velocity of propagation in the cable and tunnel and hence on the permittivity of the cable dielectric (3). Useful, practical and theoretical work was discussed at a symposium on leaky feeder radiocommunication systems, with the papers reprinted in the May 1975 issue of the *Journal of the (UK) Institution of Electronic and Radio Engineers* (4). The papers range over a wide range from reports on comparative measurements on cables to the electromagnetic theory of the leaky coaxial cable. This last is of more general interest, suggesting that a cable's capacitive coupling impedance (Z_F) has no influence on the actual screening performance of a cable but only on the measurement set-up.

B.3.3 Near Field Coupling from Cables

It has been found that in an industrial environment the principal coupling mode is from direct capacitive and mutual inductance between contiguous cables (5). The term industrial environment is used to describe a place where there is a great mass of structural steelwork and other conducting metalwork and at the same time a multitude of cables. Instrument and power cables which never run together can be coupled by this mode through the intermediary of otherwise unrelated cables. Fig. B.4 shows a sketch of a typical coupling path between a disturbing power cable and the screen of an instrument cable. Inductive coupling can take place at b and capacitive coupling at c. The coupling efficiency will be at its greatest at frequencies where one or more of the cables resonate. At current nodes the voltage will be a maximum and capacitive coupling will predominate, while at current antinodes the situation favours inductive coupling. A high coupling gain can be realised where all three cables resonate at the same frequency, the juxtapositioning is favourable for coupling, and the cable jacket is not too lossy a dielectric.

In an industrial environment, cables are normally laid on metallic trays used in close proximity to other cables. The sheath and surrounding metalwork form a transmission line with a characteristic impedance rarely in excess of 50 Ω. It is convenient to consider the coupling, not in terms of the near field for each line but in terms of the voltage and current induced in one circuit from current flow in another. Z_T and Z_F are therefore used to relate voltage inside a cable to current flow on the outside and the opposite way round. When lines are not terminated in their characteristic impedance, the relative importance of Z_T and Z_F will vary. The sheath to surrounding metalwork line comes into this category, with the effect of increasing dependence on Z_T near resonant current nodes.

B.3.4 Mixed Coupling

A radiated field incident on a metallic complex will induce currents into the individual conductors, with more current in those that resonate at the incident carrier frequency. These resonant lines will couple to other lines as described above. It is therefore possible for the sensitivity of a cable to be enhanced by the presence of others. As in Section B.3.3, it can be shown that the resonant cable need not run in close proximity to the sensitive cable for all or any part of its length, but can couple through other passive resonators.

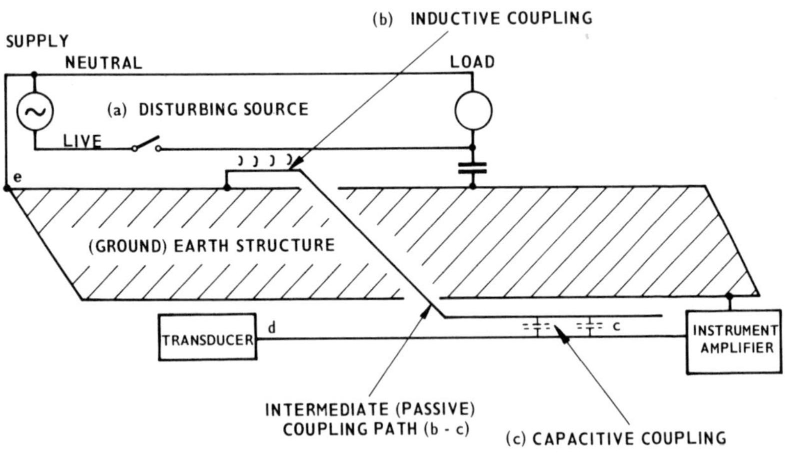

Fig. B. 4
EARTH COUPLED INTERFERENCE

B.4 Situations Needing Further Study

Problems in the areas outlined in the Introduction to this Appendix, and others that may come to mind, divide into two groups depending on the approach. This grouping is based on broadly economic considerations for those systems where the cable is the major capital item; and a rigorous theoretical and practical investigation is required. Conversely, in other systems where cables are a small part of the cost, however vital they may be, the complications of layout force the study to be of "worst-case situations."

CATV and leaky feeder systems come into the first group, and the more sensitive instrument systems form the core of the second group. It is in the multitude of medium-sensitivity instrument systems, designed without reference to external influences, that help is needed in the fields of RFI and EMC (radiofrequency interference and electromagnetic compatibility).

At present, RFI is treated by decreeing how much spurious radiofrequency voltage may appear at the terminals of an instrument (mains supply terminals are implied), often without regard to the RF fields generated by cables or from the body of the instrument itself. Similarly, many EMC studies are equally blinkered, and assumptions are made that disturbances causing malfunction are the responsibility of the user and cannot be attributed to poor screening.

No pretence is made that the problems are easy, but there is a growing need for widely accepted conversion equations relating near and far field radiated interference, conducted interference and the penetration of cable screens, connectors and screening boxes.

An attractive approach may be to decide the dominant mode of disturbance for each frequency band, spacing distance and environment (open space, metallic structure, in company with many cables etc.). Levels of disturbance can then be detailed for the dominant mode only, as is already the accepted procedure in some fields. The near field coupling from cables described in Section B.3.3 originates predominantly from 415

V/240 V 50 Hz supply cables. The predominant coupling mode from currents induced in instrument cable screens is through the cable surface transfer impedance (Z_T), and nucleonics instrumentation is required to tolerate ground coupled currents (I_o) of up to 100 mA rms over the frequency range from 10 kHz to 100 MHz (6). Even here, the relationship between I_o and instrument input may be complex, depending as it does on resonances in unterminated lines; but in most worst-case situations it is adequate to multiply the cable Z_T and length and add the transfer impedance of connectors (Z_{21}) and Z_{SH} of instrument boxes to obtain the overall screening performance of the system. It is thought that this mode is dominant in many situations, but this requires further investigation and discussion.

B.5 Proposals for Further Work

B.5.1 Measurement, supported by theory, is required to establish conversion equations from accepted far field radiation levels (in V/m) to currents flowing in a conductor and hence or otherwise to the transfer characteristics of a cable. Assume and ensure that the environment does not introduce unwanted resonances by using a single line at a fixed distance from a conducting ground. Reciprocity ensures that the opposite effect, that of radiation from a cable, will also be covered.

B.5.2 Similar expressions need to be derived for the situation where the test cable is in close proximity to a host of other cables and may be lying on a metal tray.

B.5.3 Measurements, supported by theory, are required to estimate the worst-case situation for the disturbance to a cable and screening enclosure when in close proximity to a radiating antenna. The frequency range of interest covers both the VHF and UHF bands (30 MHz to 3 GHz).

B.5.4 There is a pressing need for test equipment which is simple but effective. An investigation is required into the relative performance of a terminated pair (the tickle stick of Section B.3.2) and a more conventional aperiodic antenna for near field simulation. This should detail the power necessary to drive them to simulate the near field expected in the worst situation envisaged over the VHF and UHF frequency band and possibly down to frequencies between 300 kHz and 30 MHz.

B.5.5 Details of the dominant EMC coupling mode in various environments require investigation, especially in the lower frequency bands (30 Hz to 300 kHz) where screens of nonmagnetic material are less effective against the magnetic portion of the near field.

It is evident that several problem areas exist, and these are outlined above. Justification for studies of this sort is based on interest in the results of such work. It is also reasonable to look for encouragement if not the expertise from organisations with such problems.

References

(1) Fowler, E.P., "A comparison of the screening performance of braided coaxial and triaxial cables," AEEW-R 982, March 1975.

(2) Simons, K.A., "Attempts to relate transfer impedance to radiation under somewhat realistic conditions," IEC Working Group paper SC46A/WG1(Simons)3, September 1974.

(3) Martin, D.J.R., "A general study of the leaky feeder principle," *The Radio and Electronic Engineer*, Vol. 45, No. 5, 205-214, May 1975.

(4) "Leaky feeder radio communications systems," *The Radio and Electronic Engineer,* Vol. 45, No. 5, May 1975.

(5) Wilson, I., *et al.*, "Electrical interference in instrumentation systems" (to be published).

(6) Fowler, E.P., "Interference immunity tests for nucleonic instrumentation," AEEW-R 919, H.M.S.O., April 1974.